Neural Interfacing: Forging the Human-Machine Connection

Neural Interfacing: Forging the Human-Machine Connection

Susanne D. Coates

ISBN: 978-3-031-00512-1   paperback
ISBN: 978-3-031-01640-0   ebook

DOI 10.1007/978-3-031-01640-0

A Publication in the Springer series
SYNTHESIS LECTURES ON BIOMEDICAL ENGINEERING

Lecture #22
Series Editor: John D. Enderle, University of Connecticut

Series ISSN
Synthesis Lectures on Biomedical Engineering
Print 1930-0328   Electronic 1930-0336

Synthesis Lectures on Biomedical Engineering

Editor
John D. Enderle, *University of Connecticut*

Introduction to Statistics for Biomedical Engineers
Kristina M. Ropella
2007

Capstone Design Courses: Producing Industry-Ready Biomedical Engineers
Jay R. Goldberg
2007

BioNanotechnology
Elisabeth S. Papazoglou and Aravind Parthasarathy
2007

Bioinstrumentation
John D. Enderle
2006

Fundamentals of Respiratory Sounds and Analysis
Zahra Moussavi
2006

Advanced Probability Theory for Biomedical Engineers
John D. Enderle, David C. Farden, and Daniel J. Krause
2006

Intermediate Probability Theory for Biomedical Engineers
John D. Enderle, David C. Farden, and Daniel J. Krause
2006

Basic Probability Theory for Biomedical Engineers
John D. Enderle, David C. Farden, and Daniel J. Krause
2006

Sensory Organ Replacement and Repair
Gerald E. Miller
2006

Artificial Organs
Gerald E. Miller
2006

Signal Processing of Random Physiological Signals
Charles S. Lessard
2006

Image and Signal Processing for Networked E-Health Applications
Ilias G. Maglogiannis, Kostas Karpouzis, and Manolis Wallace
2006

Neural Interfacing:
Forging the Human-Machine
Connection

Susanne D. Coates

SYNTHESIS LECTURES ON BIOMEDICAL ENGINEERING #22

ABSTRACT

In the past 50 years there has been an explosion of interest in the development of technologies whose end goal is to connect the human brain and/or nervous system directly to computers. Once the subject of science fiction, the technologies necessary to accomplish this goal are rapidly becoming reality. In laboratories around the globe, research is being undertaken to restore function to the physically disabled, to replace areas of the brain damaged by disease or trauma and to augment human abilities. Building neural interfaces and neuro-prosthetics relies on a diverse array of disciplines such as neuroscience, engineering, medicine and microfabrication just to name a few. This book presents a short history of neural interfacing (N.I.) research and introduces the reader to some of the current efforts to develop neural prostheses. The book is intended as an introduction for the college freshman or others wishing to learn more about the field. A resource guide is included for students along with a list of laboratories conducting N.I. research and universities with N.I. related tracks of study.

KEYWORDS

Neural Interface, Neuro-prosthetics, Neural Prosthetics, Neural Engineering

Contents

For all my mentors: past, present, and future.

Art, Science, and Metaphysics

Understanding the human body is certainly one of the most daunting challenges facing science today. Although we tend to think of this undertaking in modern terms, it is not a new endeavor. Since the beginning of recorded history, we have been trying to understand ourselves from both a physiological and metaphysical standpoint. That is, what the different organs of the body do, what causes disease, what is intellect, where is the seat of "consciousness," and is reality "real." There have been many people both ancient and modern who have contributed to the body of knowledge: Plato, Hypocrites, Galen, Descartes, and Ramón y Cajal are a few that immediately come to mind. However, the advancement of human thought and knowledge has not been a linear process. There have been many false starts, periods of enlightenment, and periods of darkness where the quest for truth (scientific or otherwise) was nearly snuffed out. For millennia we have been asking the question "who and what are we?", but it is only in the last few centuries that we have begun to get some "real" answers.

Our ability to understand the world around us is bound by the limitations of our brain and five senses. While we know the limitations of our five senses, we cannot say with any great certainty what the limits of the brain are, especially with respect to cognition. Thus, it is difficult to ascertain the ultimate limits of our ability to understand. The advances in human knowledge that have occurred in the last several hundred years have happened for a number of reasons. Chief among these is the de-mystification (or perhaps the re-demystification following the middle ages) of the natural world. That is, moving away from the belief that nature, natural processes, and the universe are mysterious and cannot be understood by man. This was a vitally important step because our perceptions of the natural world and our will to explore it can be greatly influenced by our beliefs.

The ability to extend the range of our senses with tools such as the telescope, microscope, X-ray machines, and others has played an important part in the discovery and elucidation of natural processes. It is through our ability to create tools that we overcome the inherent limitations of our bodies. Although neural interfaces and neuro-prosthetics may initially be used for rehabilitation and restoration of function to the disabled, it is probably inevitable that they will become the means by which we will further extend our senses and augment our perceptual abilities. How might the human ability to understand the world around us be changed should it one day be possible to cybernetically augment our brains and radically alter our perceptual experience? How we view the universe and our relationship to it is important because it is through this lens that we interpret what we experience. Our knowledge, cultural precepts, religious beliefs, and personal philosophies both influence and are influenced by our day-to-day interactions with our environment and with each other.

As an example of how one's personal philosophy may influence how one interprets the world around them, consider this passage from the Tao Te Ching[1]:

---

[1]Tao Te Ching by Lao Tzu, p. 11, Translation by Gia Fu Sheng and Jane English.

Thirty spokes share the wheels hub;

It is the center hole that makes it useful.

Shape clay into a vessel;

It is the space within that makes it useful.

Cut doors and windows for a room;

It is the holes which make it useful.

Therefore profit comes from what is there;

Usefulness from what is not there.

Thus, where one person might see a room filled with furniture as a space defined by the objects it contains, a Taoist might see the same room full of objects as being defined by the spaces between the furniture. This may seem like a trivial distinction, however, it is germane to our discussion for it represents an alternate point of view applicable to a number of problems. For instance, in the following figure do you see a series of vertical lines occurring in a given space of time or a series of spaces of various size that are delineated by vertical lines?

In this case, these two different points of view represent two different ways of looking at the issue of neural coding. That is, understanding the code used by brain cells to communicate with one another. The first point of view, the number of vertical lines (or spikes) occurring in a given amount of time, represents what is known as a rate code. The second point of view represents temporal coding which looks at the change in the intervals between the spikes. Thus, two different ways of looking at the world (spaces *vs.* objects) can translate into two different but equally relevant interpretations of the same data.

Science may seem the antithesis of art and other creative activities due to its heavy focus on logic and reasoning. This emphasis on logic and reason does not mean that there is no room for creativity in the sciences; in fact, creativity is essential. Regardless of your field of study, creative thinking is necessary for the generation of novel ideas and new points of view. In science, producing new techniques, new ways of looking at old problems, and new avenues of discovery may require that the person free themselves from conventional ways of thinking. That's not to say that one should ignore an existing body of work, but one should not be unnecessarily bound by it either. Creativity and thinking "out-of-the-box" is as crucial in the sciences as it is in any other discipline, although not as widely recognized as being so. Scientists often get stereotyped in pop-culture as the absent-minded professor, the arrogant genius, or the anti-social geek. These stereotypes are caricatures that do not reflect the wide range of individual variation. In my time in science I have met many diverse, talented people who do not fit this popular mold. Really, the *only* thing you can say of people in the sciences as a group is that they tend to be creative, intelligent, multi-talented individuals with a passion for discovery and invention.

In both modern and ancient culture science sometimes finds itself at odds with religion. In a politically charged environment where issues become rapidly polarized and the middle ground falls away, civil discourse is of paramount importance. Civilized society depends upon, at the bare minimum, tolerance for points of view different from ones own. Ideally, understanding and respect are preferable to mere tolerance. The intersection of science and religion is often a contentious one. On issues such as stem cell research, there are many opposing points of view and those who are very passionate in their assertions of what is morally and ethically correct. In such a milieu it is very difficult to achieve compromise. The goal of building a culturally diverse, civilized global society is not possible without the ability to find common ground, but to find something one must have the desire to look for it first. Compromise is not possible without a willingness to listen to and respect other points of view.

My interests tend to gravitate toward the center of the Art-Science-Metaphysics triad in an effort to achieve a balance in my own life. I believe that in the larger sense these three disciplines represent three different ways of exploring the same question: "What is the nature of existence?" Unfortunately, a harmony between these three disciplines at a societal level has often been problematic due to their vastly different ideologies. Science is primarily interested in the logically quantifiable aspects of characterizing natural processes through observation, formulation of testable hypotheses and experimentation. Art, a more difficult to define endeavor, generally focuses on communication of the internalized emotional and/or spiritual aspects of the artist's world. Metaphysics, with its two major subdivisions ontology and theology, ponders deeper and possibly un-answerable questions such as "Is there reality separate from mind?" and "What is the purpose of the universe and what is man's place in it?" To sum this up in a sound-bite:

Art, Inspires

Science, Explains

Metaphysics, Cogitates

Regardless of your religion, philosophy of life, or where your personal journey leads you, we all have something to contribute. It is as a collective that humanity will move forward and advance our understanding of both ourselves and the world in which we live. The specifics of how each of us proceeds in our own personal search for meaning in our lives may be very different. However, the underlying reason of *why* we do it is not. We are all on the same quest to try to understand our place in the universe. In that respect, we are not so different after all.

Susanne Coates, August 8, 2008

CHAPTER 1

Introduction

"All our knowledge begins with the senses, proceeds then to the understanding and ends with reason. There is nothing higher than reason."

- Immanuel Kant -

From the lost civilization of the Krel in Cyril Hume's 1956 movie *Forbidden Planet*, to the cybernetically enslaved humans of Andy and Larry Wachowski's *The Matrix*, writers of science fiction have long imagined the interfacing of humans and machines. While machines (most notably computers) are often used as analogies for how the brain functions, no two systems could be more different. Machines, assembled from hard-wired component parts, rely on software to endow them with the flexibility to accomplish different processing tasks. Humans, on the other hand, reach their completed, adult state through a process of growth and development. Beginning in the womb and continuing until a few minutes after we draw our last breath, the nervous system is always on and always changing in response to a continual stream of sensory information. Thus, as we experience and learn, subtle changes in the brain incorporate this new information into the existing structure. A connection might be slightly altered here, a shift in the amount of a particular neurochemical there, minute changes all to preserve, for example, the sight, sounds, smells, and emotional response of seeing an elephant for the first time. While computers are lousy at writing a good poem about that elephant, they are very good at calculating the elephant's internal volume or analyzing elephant population statistics. The differences between biological neural systems and digital computers give rise to strengths and weaknesses unique to each. The phrase "brain-machine interface" (BMI) brings to mind the futuristic idea of seamlessly integrating computers and the brain for the purpose of augmenting human abilities. Presumably, such a human would be able to capitalize on the strengths of both systems. While this may someday be possible, there is a vitally important and immediate application of this technology.

Traumatic injury and diseases such as Parkinson's, Huntington's, and Alzheimer's disable millions of people every year, resulting in the loss of physical and/or mental function. The primary goal of neural interfacing and neuro-prosthetics research is to develop implantable systems to restore function to the mentally and physically disabled. Although still in its early stages, this research holds great promise for helping the blind to see again, paraplegics to walk, and quadriplegics to interact with the world using computers in ways not possible a few years ago [1]. One of the keys to achieving the lofty goal of restoration/augmentation of function using artificial means is the interface; the human-to-machine connection.

This first chapter is intended to introduce some of the terminology, neural physiology, and other background material germane to the discussion of neural interfaces. If you desire a more detailed discussion of the topics touched upon in this chapter, please refer to the numbered references in the bibliography. (If you would like to delve deeper into neuroscience or neurophysiology related topics, I suggest beginning with the latest edition of "From Neuron to Brain" by Nicholls et al. [2].) Chapters 2 and 3 explore interfacing methods and discuss some current neuroprosthetic applications. Chapter 4 speculates on some of the more far-reaching applications of neural interfacing (NI) and touches on ethical issues. Lastly, Chapter 5 provides information for students interested in pursuing a career in neural engineering.

1.1   THE ROAD AHEAD

One of the greatest challenges facing neuroscientists today is understanding in great detail how the brain works. In other words, how the structure, physiology and biochemistry of the brain relates to its high level function. Remarkable progress has been made in this area but there is still much that we do not understand. For example, is memory encoded as structural features, patterns of gene activation, as patterns of activation across the neural network, in the pattern of the electrical field itself or all (or none) of the above? We have identified many possible mechanisms that play a role in memory and other functions. But how these mechanisms fit into the bigger picture of how a specific memory is stored still eludes us. Fortunately, a highly detailed understanding may not be necessary to, for example, connect a fully functional prosthetic arm because there are areas of the brain where the neural activity correlates well with limb movement. Thus, by placing sensors in these areas and performing suitable processing on the acquired data, the user can direct the motions of a prosthetic limb or the cursor on a computer screen [1]. Unfortunately, there are few such correlates for memory and even fewer for cognition. If we eventually hope to build replacement parts for damaged brains, augment intelligence or integrate vast computer databases with human memory we must have a far greater understanding of the brain than we have at present. Advancing our understanding of the brain is pivotal to the continued advancement of neural interfaces and neuro-prosthetics.

Historically, the key to increasing our understanding has been the development of new and better tools with which to study the brain. The desire to comprehend our surroundings either on very small (e.g., the sub-atomic) or very large (e.g., the universe) scales and our lack of a natural means to do so, underscores the importance of tools and technology. While it is often the case that the development of new tools precedes major discoveries, it is also true that those same discoveries may further the development of the next generation of tools. In science there is not a clear division between tool-building and tool-using activities since scientists have typically been both the builders and users. This is a role that has arisen out of necessity because the needed tools often don't exist (try finding a particle accelerator at your local hardware store) and it's up to the researchers to create them to facilitate their work. Simply stated, tools are the means by which we bring into our conscious perception the processes occurring at normally imperceptible temporal and/or spacial scales; without them, there is little hope of elucidating mysteries of the brain. Perhaps I am biased since I spend much of my time designing and building tools for others to use, but I believe that there is something primal, something uniquely human (or at least uniquely primate) surrounding the activity of making and using tools. After all, it *is* one of the key things that sets us apart from nearly all other animals on this planet.

How information processing and storage might be accomplished in the brain is a topic to which whole books have been dedicated. Research continues to uncover new information which sometimes forces us to re-evaluate our theories on brain function. Unlike computers, in the brain there is no distinction between software and hardware. The physical, chemical, molecular, and perhaps even quantum [3, 4] structure of the brain alters itself from moment to moment to store and/or process information. It is not known how this process is directed. What are the underlying rules (if any) that dictate the proper way to modify structure to accomplish the given processing task? To use the computer analogy, what we are looking for is the brain's operating system. Realistically, what we are finding is something very different. In fact, it is almost certain that no two brains process the same information in exactly the same way.

There *are* commonalities between all normal human brains such as cytoarchitecture, regional organization or the organization of fiber tracts (see below), but the commonality apparent at the gross anatomical level may not be relevant to determining how information processing and storage is accomplished in an individual brain. For example, vision processing is localized to specific regions of all brains; however, the specific way a given visual experience is perceived, understood, and integrated into memory by a given individual is unique. It's as if each brain develops its own unique operational schema (conceptual framework) within the bounds common to all human brains. So, while it is possible to place an electrode in the brain that will receive a signal whenever you move your right index finger, it is not possible to do the same for recalling the experience of listening to your grandmother tell you a bedtime story. This is because episodic memory (memory of places, events, emotions, etc.) results in a pattern of activation across many brain regions. Using tools which allow visualization of whole-brain activity, it may be possible to teach a system to recognize the unique pattern of activation that corresponds to a mental state or (in the future) an already stored experience [5]. The inverse operation of predicting the unique pattern of brain activation that will result from a given experience or being able to look at brain activity to tell what someone is thinking, may not be possible for a very long time to come.

Computers are often used as a convenient analogy for describing how the brain operates; in actuality, digital computers and the brain have little in common. Whether you are a fly or a human your brain makes use of massively paralleled architecture, electro-chemical communication, genetic regulation, and a host of other complex biological processes. The human brain and nervous system make decisions about a continual stream of information, coordinate responses to that information, manage all biological functions, and are responsible for the epiphenomena we refer to as self awareness. Computers, on the other hand, operate by performing a set of pre-determined, sequential operations on an input to produce an output. While modern computers permit certain types of parallel processing they are not truly parallel. They simply break a larger problem into smaller chunks and operate on the chunks semi-independently (and sequentially) often waiting for other parts of the program to finish before the next series of parallel processing operations can continue. Only certain types of problems can be solved in this manner, so the parallel processing strategies of the present are not even generalizable to all types of problems that can be solved on a computer.

In terms of raw number crunching ability, a handheld calculator can far outmatch most humans in a contest of speed at performing complex mathematical operations. However, the human contestant with their paper and pencil is actually performing a far more complicated feat. The brain is processing various sensory input streams, calculating how hard to press the pencil to the paper, controlling the muscles of the arm and hand, maintaining posture, using language to create the characters that are written on the paper, and controlling a host of other necessary functions that are all completely transparent to the human contestant scribbling away on the paper. While our brain was not specifically "designed" for number crunching, such computation is one of a great number of tasks that it can perform.

The chess playing computer Deep Blue that outmatched world champion Gary Kasparov in 1997 was heralded by some in the artificial intelligence community as a sign that machines which rival the human brain in ability were not far away. However, Kasparov was defeated not by a "thinking" synthetic opponent but by custom hardware designed *only* to play chess and with enough speed that it could evaluate over 200 million positions per second. In short, Kasparov was defeated by brute force rather than intelligence. I would also like to point out that following the match, Gary Kasparov with his "slower" human brain was able to stand, walk from the building, and continue with his life. Deep Blue, on the other hand, was powered down, disassembled, and is now on display in a museum. Why? Because despite its parallel processors, custom chips, and clever software it was useful for no other purpose than defeating the chess master. Thus, Deep Blue was more a testament to its designers engineering ability and determination to build the "ultimate" chess-playing computer than a milestone on the road to adaptable, "intelligent" machines.

1.2   THE BRAIN AND NERVOUS SYSTEM, IN A NUTSHELL

The brain is the organ of thought just as the heart is the organ of blood circulation. This statement may seem obvious but the idea is of relatively recent origin. Beginning with the ancient Egyptians and well into the Middle Ages, it was the heart that was believed to be the organ of thought. The brain was a minor organ that filled up ones skull or became infested with daemons that had to be released through trepaning (drilling holes in ones head). Although the ancient Egyptians ascribed only minor importance to the brain, they were the first to begin describing its anatomy[1]. Later, in 450 B.C., the Greek physician Alcmaeon of Croton [6], based on his dissections of animals, was the first to suggest that it is the brain, not the heart that is the seat of intelligence. This notion was accepted by some such as Plato but rejected by Aristotle who contended that the brain is a radiator which cools the blood. By 300 B.C., the prevailing opinion had shifted to the ventricles of the brain being the seat of intelligence. While still incorrect, at least it was in the right general area. The Middle Ages saw a complete halt to brain research because of the church's ban on the study of anatomy. It wasn't until the 1600's that the French philosopher Reneé Descartes[2] properly identified the brain as the seat of intellect and 18 years later that Thomas Willis, M.D. at Oxford wrote the first neuroanatomy text[3]. The Willis text is considered the beginning of the modern descriptions of brain anatomy.

There are different levels to the structure/organization of the brain, including: gross anatomical (Fig. 1.1), cytoarchitectural, and functional. "Gross anatomical" refers to the unique features of the brain and nervous system that are observable with the naked eye. This might also include visible connections between different brain regions or between the brain and the parts of the body to which it connects. This was the first type of description we have of the brain since early anatomists did not have tools like the microscope and staining techniques to study it in greater detail. It wasn't until the time of Camillo Golgi and Santiago Ramón y Cajal (mid to late 1800's) that microscopes had become sufficiently powerful and stains were developed which permitted the visualization of the fine structure of the brain. Gross anatomy tells us little about what the various structures do, but it does provide landmarks to aid us in finding our way around the landscape of the brain. The classic, gross anatomical reference for the brain and body, *Anatomy Descriptive and Surgical*, was first published by Henry Gray in 1858.

Histology is a branch of anatomy which studies the minute structure of animal and plant tissues. Histology, as a field of scientific inquiry, began with the invention and refinement of the light microscope. Cytoarchitecture refers to the use of histological features to differentiate brain regions from one another based on cell composition and arrangement. At the beginning of the 20th century, Korbinian Broadmann used cytoarchitectonic studies performed by himself and others to divide the brain into 52 regions referred to as Broadmann areas [2, p. 364]. Some Broadmann areas (BA) were later found to correspond to regions that were identified during functional mapping of the brain. For example, the somatic sensory map (Fig. 1.2) corresponds to areas 3, 1, and 2, and the primary motor map (Fig. 1.2) is located in BA 4.

Initial descriptions of the function of the various regions were based on studies of brains damaged by injury and disease which led to, for example, aphasia or loss of the ability to use and/or comprehend words [8, pp. 391-393]. Also, electrical probing of the cortex (the surface of the brain) by researchers in the 1800's revealed areas which seemed to play a role in specific functions like movement and vision. These methods, though useful for early descriptions of functional localization, were fraught with uncertainties. In the case of brain injury, for example, was the damaged area the locus of the lost functionality or did the injury simply damage connections passing through that area on the way to another area of the brain?

[1]The Edwin Smith Surgical Papyrus.
[2]Les passions de l'ame, 1649.
[3]Cerebri Anatome, 1664.

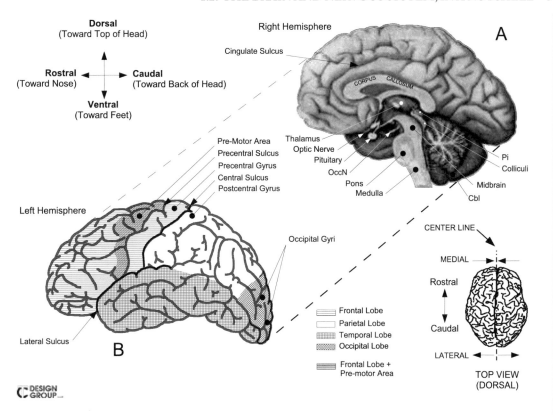

Figure 1.1: Illustration showing some of the gross anatomy of (A) the medial aspect (sectioned along the center line) of the right cerebral hemisphere, midbrain, and brain stem. OccN: Occulomotor Nerve; Cbl: Cerebellum; Pi: Pineal Body (adapted from [7, Fig. 715]). (B) Drawing of the left cerebral hemisphere of the human brain showing the approximate locations of the lobes (based on [7, Figs. 728 and 744]).

More recent methods using electrical/chemical stimulation and recording via microscopic electrode arrays have helped to form a better picture of the brain's functional localization and connectivity.

One of the best known examples of a functional map is the sensory-motor representation located on either side of the central sulcus on both hemispheres of the brain (Fig. 1.2). The neurons of corresponding areas of the motor map (on the pre-central gyrus, Fig. 1.1) are active during volitional movement of the given part of the body, while the neurons of the sensory map (on the post central gyrus, Fig. 1.1) are active when sensory input (pressure, hot/cold, etc.) is received from the corresponding area of the body. Most of the areas represented on the sensory-motor map in the left cerebral hemisphere are connected to the right side of the body and vice versa. Referring again to Figure 1.1-B, we see another area of particular interest in neuro-prosthetic limb research, the pre-motor area. This area is involved in movement planning and encodes information about joint angles, limb trajectory and other information useful for command and control of a prosthetic limb.

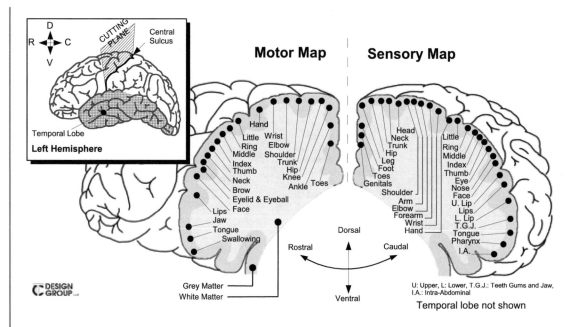

Figure 1.2: Diagram of the motor and sensory representations in the left hemisphere of the brain. Note: Eye, Nose, and Tongue on the sensory map refer to sensations of heat, pressure, etc. Vision, olfaction, and taste processing are located in other sense specific brain areas.

The human brain receives sensory input from all parts of the body and exerts control over all bodily systems both voluntary and autonomic. The bulk of the connections to and from the brain are made by nerve fibers that traverse the spinal cord. Injury to the spinal cord interrupts these connections causing varying degrees of functional loss below the site of the damage. The spinal cord is more than just a conduit for these fibers. Like the brain, it has white matter composed of fibers and gray matter containing the neuron cell bodies and their local interconnections. Unlike the brain, the spinal cord's white matter is located at the periphery while the gray matter is centrally located (compare Figs. 1.2 and 1.3). The cord contains neurons involved in, for example, automatic reflexes responsible for coordinated walking/running, flexor withdrawal (withdrawing your hand from a hot surface), bladder control, and sexual function. Efferent fibers traveling from the cord to the periphery have their cell bodies in the gray matter of the cord. Afferent fibers from the somatic sensory system have their cell bodies in the dorsal root ganglia (see Fig. 1.3). In addition to their connections with the brain, both efferent, and afferent fibers make synapses with other spinal neurons to enable the previously mentioned reflex responses. The projections from spinal neurons enter/exit the spinal column via the dorsal/ventral horns located on either side of each segment of the spinal cord. The spinal nerves are formed by the merging of fibers from the dorsal root ganglia and ventral horn (Fig. 1.3). Spinal nerves branch and merge many times along their path in the body as groups of fibers split off to connect many of the body's systems to the brain. Some of these fibers, such as those connecting the legs to the spinal cord, can be over a meter in length. Like the brain, the spinal cord is highly organized and this organization can be seen at both the gross and microscopic level (Fig. 1.3).

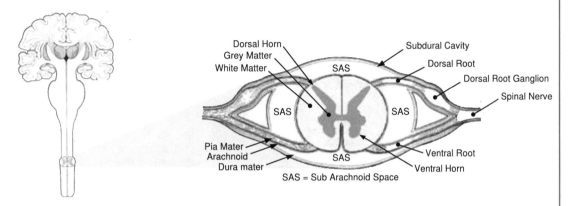

Figure 1.3: Cross-section of the spinal cord. Adapted from [7, Figs. 759 and 770].

The entire nervous system functions as a unified whole, blissfully unaware of any divisions that we might assign. In essence, this is much like the lines drawn on a map to delineate the borders of countries, states, and cities; the divisions are largely for our convenience in discussing the brain, spinal cord, and all of the attached systems.

1.2.1   NEURONS

The brain and nervous system are composed of trillions of cells and there are many ways to classify the different type of cells which comprise this total. For the purpose of our discussion, it is sufficient to say that there are many different types of cells and each cell type plays a unique role (either directly or indirectly) in information processing. The major cell type that is directly involved in information processing is the neuron and there are approximately 100 billion of these in the adult brain. Neurons communicate with each other through their axons and dendrites (Fig. 1.4-A). In general, dendrites conduct impulses toward the cell body (soma) while axons conduct impulses away from the soma (Fig. 1.4-A). It has been estimated that in the adult brain there are approximately 4–6 quadrillion interconnections between neurons alone. The actual number of interconnected cells is far greater when you consider the neuroglia which also have a role in information processing in the brain.

A *synapse* is the point of contact and information exchange between neurons. The brain "processes" approximately 1 quadrillion (1×10^{15}) synaptic events per second. These events are not confined to the simple relaying of messages from one side of the synapse to the other; rather, complicated chemical, cellular, molecular, and even genetic mechanisms can modify the message at the synapse in ways only the brain understands. Any comparisons between the "processing" speed of the brain and that of computers is purely a guess since (1) the two systems are so different and (2) it is presently not possible to know all the processing-relevant cellular/molecular events in the brain that are taking place in a given interval of time. Even considering only the number of synaptic events per second, the brain far exceeds even the fastest modern computer.

A synapse is composed of a presynaptic membrane (on an axon), the synaptic space, and the post-synaptic membrane (on a dendrite). Dendrites may be irregular or "bumpy" in appearance due to the presence of synaptic "spine." The spine is formed due to varicosities along the length of the axon where they synapse with dendrites. The dendrite in turn develops a warp in the post-synaptic membrane to "clasp" the axonic varicosity. This gives the appearance of a bump on the dendrite since the smaller portions of the axon are not as easily visualized as the larger varicosity. The term "neurite" is used to refer to either an axon or a dendrite while the plural "neurites" may refer to both simultaneously. When an electrical impulse (termed an action potential or AP) arrives at the pre-synaptic membrane of a chemical synapse it triggers the release of small molecules (neurotransmitters) which cross the synaptic space, bind to receptors on the post-synaptic membrane (Fig. 1.4-C), and either increase or decrease the probability that an action potential will be generated at the axon hillock (Fig. 1.4-A) of the post-synaptic neuron.

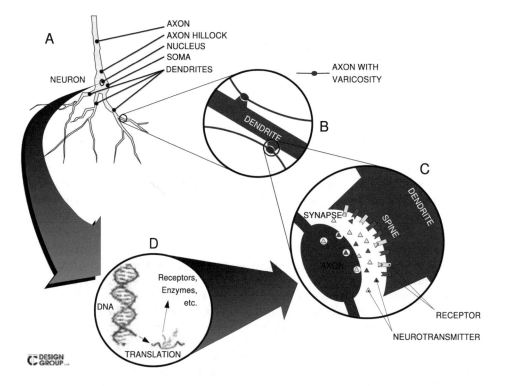

Figure 1.4: Diagrammatic representation of complexity at the level of the neuron: (A) The neuron shown above has a dendritic tree formed by three main branches projecting from the bottom of the cell and sends out a single axon from the apex. The axon hillock is the junction of the soma and axon. (B) Dendrites are studded with synaptic spines and receives information from other neurons. At each synapse, (C) neurotransmitters are released from the pre-synaptic membrane and bind receptors on the post synaptic membrane. (D) Within the somas of pre- and post-synaptic cells, genes that code for receptors and neurotransmitters may be up or down regulated in response to activity or other events.

Neuroglia, while long thought to only provide metabolic support to neurons, are now known to be involved in synaptic transmission and information processing. Astrocytes respond to the level of synaptic activity and exhibit reciprocal communication with neurons via calcium signaling [9]. It has long been known that both neurons and neuroglia can produce neuromodulatory substances which effect the probability that activity at proximally located synapses will result in the generation of action potentials. The effect neuromodulation has on information processing and the exact mechanisms whereby it is accomplished are still the subject of investigation.

Every living cell has a charge difference across its plasma membrane due to differences in the concentration of ions between the inside and outside of the cell. Neurons contain specific cellular mechanisms to manipulate this charge differential to accomplish signaling between cells. In a neuron in its resting state (not conducting a signal) the inside of the cell is about -70 millivolts with respect to the outside. This potential difference, measured across the plasma membrane, is created largely due to a higher concentration of cations (positively charges ions) in the extracellular fluid and a higher concentration of anions (negatively charges ions) in the intracellular fluid (cytoplasm). On the outside, the principle ions are Sodium (Na^+) and Chloride (Cl^-) with Potassium (K^+) present in low concentration. Inside the cell the situation is reversed: (K^+) is the principle cation while (Na^+) and Chloride (Cl^-) are present in low concentration. The potential difference is created by the presence of anions in the form of proteins, amino acids, phosphate, sulfates, etc., in the cytoplasm [10].

Cells are leaky; sodium and potassium constantly diffuse across the plasma membrane following the concentration gradient. That is, without a mechanism to constantly maintain the ion (charge) gradient the neuron would not have a strong potential difference across its plasma membrane. Without this potential neurons would be unable to propagate action potentials and thus unable to convey information. To maintain the ion gradient Na^+/K^+ pumps are interspersed along with other cellular apparatus throughout the plasma membrane. These pumps are constantly moving Na^+ ion from the inside of the cell to the extracellular space and moving K^+ the opposite direction. In addition to diffusion the membrane is also selectively permeable to certain ions. This selective permeability and its ability to be modulated by both internal and external influences is the basis for information processing and signaling in neurons.

Action potentials (a.k.a. spikes) are generated at the axon hillock in response to the summated electrical activity occurring at all of the neuron's synapses (Fig. 1.4-A). When this summated activity results in the membrane potential at the axon hillock shifting positive from -70 mV (resting) to about -50 mV (threshold potential), voltage gated ion channels open allowing a mass influx of sodium ion into the cell. This decreases the potential difference between the inside and outside and is termed depolarization. An above threshold depolarization in one area of the axon causes adjacent channels to open which depolarizes the next area and so on in a cascade effect which results in action potential propagation down the axon. When the depolarized area has reached about +35mV, the Na^+ channels have closed and the K^+ channels have opened allowing the flow of positive charge out of the cell to temporarily restore the local resting potential. Since the K^+ channels close slowly this results in undershoot and slight hyper-polarization (Fig. 1.6). After a few milliseconds the Na^+/K^+ pumps have restored the ionic concentrations to their resting values. During the re-polarization phase and undershoot the neuron is not capable of generating an AP in response to stimulus. This time interval is referred to as the refractory period. The refractory period places a fundamental limit on the maximum rate of firing of a given neuron.

Trains of action potentials manifest as moving waves of focal depolarization down the axon. The spikes you see on the oscilloscope result from the flow of electric current created by the movement of ions as these waves pass the stationary electrode. Multiple spikes are referred to as a *spike train* and when a neuron is producing spikes we say that it is "firing." Depending on the location of the electrode you will see the depolarization as a positive- or negative-going event (Fig. 1.5). That is, if your electrode is on the inside of the cell you will see the resting potential at -70 mV and the peak of depolarization at approximately +35mV just as discussed above.

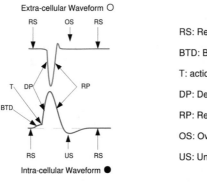

Figure 1.5: Comparison of extracellular and intracellular waveforms.

The values of potentials, shapes of waveforms, etc., are based on the study of neurons using intracellular recording techniques. All of the neural interfacing applications that I am aware of at the time of this writing, use extracellular electrodes which are located in the fluid filled space outside the cell. The waveform recorded by such electrodes will be inverted and its shape/amplitude will be altered as the distance from the source increases. The extracellular and intracellular waveforms are not mirror images of one another. The extracellular waveform is the result of the bulk current flow created by the movement of ions in the vicinity of the electrode. As such, details like the below threshold depolarization and undershoot which are visible using intracellular recording techniques are not visible in the activity recorded by an extracellular electrode (Fig. 1.5).

Action potentials are generated in an "all-or-none" fashion. This means that once the stimulus is sufficient to trigger an AP, a greater intensity of stimulus does not equate to a larger amplitude spike. Rather, the intensity of the stimulus is encoded by the relative timing characteristics of the individual spikes comprising a spike train. It is a popular mis-conception to think that neurons communicate like components in computers, using "digital" signals. This probably arose from a mis-understanding of the all-or-none characteristics of AP propagation and the fact that at first glance spike trains somewhat resemble digital pulse trains. However, if you are searching for an analogy, analog frequency modulation (FM) is probably a better (though not entirely accurate) one for the way neurons communicate with one another.

Communication between neurons occurs through a combination of electrical and chemical signal transduction. The reason for this is relatively straightforward, the all-or-none method of electrical transmission is well suited for rapid signaling over distance but lacks a robust method for modification of the message. Chemical signaling permits robust information processing at the level of the synapse through alteration of the message by amplification, attenuation or modulation of its content. For example, when an action potential arrives at the pre-synaptic membrane a quantity of neurotransmitter (NT) is released. The amount of NT released could be dictated by genetic factors (i.e., up- or down-regulation of the gene for that particular NT thus altering the amount available for release), the number of action potentials arriving at the pre-synaptic membrane during a given time, (i.e., high-frequency activity might cause more release than low-frequency activity), the speed with which NT is made available for release, etc. Once the NT has been released it will diffuse across the synaptic space and bind the receptors on the post

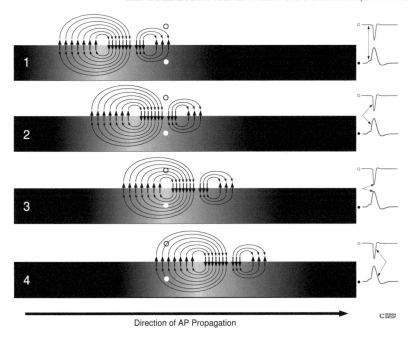

Figure 1.6: Diagrammatic representation of action potential (AP) propagation along a small section of an axon and its relationship to the recorded waveform. The "wave" of depolarization is depicted as a lighter area moving along the axon. Current flow is shown as circular arrows. Note that depending on whether the recording electrode is located inside (●) or outside (○) the cell the detected waveform will be inverted. (1) An AP arrives from the left and triggers the opening of voltage gated ion channels (VGIC's) which permit the influx of positive ions (Na^+) into the lumen. This depolarizes the membrane forward of the site of opened VGIC's (smaller circular arrows). When the depolarization reaches threshold the VGIC's open, the process begins again and the AP continues to propagate down the axon. Due to the the refractory period, the AP only propagates in one direction (to the right) as seen in 2–4.

synaptic membrane. These receptors open ion channels which can either hyperpolarize the cell (making it less likely to fire) or depolarize the cell sending the axon hillock closer to the -50 mV threshold needed for the generation of an action potential. Thus, the number and type of receptors (as well as other factors) influence the probability that the post-synaptic neuron will generate an action potential in response to the pre-synaptic activity.

In Figure 1.7 we see a simple example of how a neural message might be modified by modulation at chemical synapses. *Attenuation* (Fig. 1.7, left) results when the number of spikes produced at the post-synaptic neuron is less than the corresponding activity at its pre-synaptic inputs. In other words, the strength of the elicited response is less than that of the stimuli. *Amplification* (Fig. 1.7, right) refers to the opposite situation where the post-synaptic response is greater than the pre-synaptic stimuli. Furthermore, the post-synaptic response can be temporally altered, that is, the inter-spike intervals (ISI) in the response do not necessarily have to be a fixed ratio of the ISIs of the stimulus.

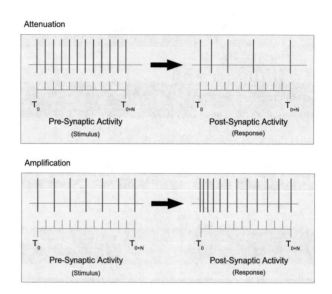

Figure 1.7: Example of attenuation and amplification of a message.

There are over 40 known types of neuro-active substances and even more types of receptors. Some of these neurotransmitters are listed in Table 1.1 along with their receptors and where they are found in the nervous system. Although the basis for chemical transmission was proposed by John Langley nearly 100 years ago, we are still discovering new substances involved in neurotransmission. For example, nitric oxide whose function as a neurotransmitter was still debated when I was in graduate school 8 years ago, has been found in many brain areas and is believed to be involved in learning and memory in the hippocampus[4]. Other molecules such as Oxytocin have dual functionality as both a hormone and a neurotransmitter.

It is likely that there are many more neuroactive substances awaiting discovery. Moreover, continued research is necessary to gain a clear picture of role that each of the *known* substances plays in brain function. The key to understanding the molecular and chemical basis of neural processing is better tools. One example of such tools is the present work being conducted to build Micro-ElectroMechanical Systems (MEMS) capable of delivering drugs, neurotransmitters, or other substances to highly focused areas within the nervous system. MEMS push the envelope of our ability to fabricate very complex systems on very small scales. Continued development and refinement of MEMS is linked to the continued advance of microfabrication technology. A system capable of delivering controlled amounts of different substances *in vivo*, with high spatial accuracy and the ability to monitor the electrical/chemical activity of the surrounding neurons would be an invaluable tool for discerning the role of those substances in neural processing.

---

[4]An area in the brain involved in memory.

Table 1.1: Examples of some currently known eurochemicals and their receptors

Name	Receptor	Location*
Acetylcholine	Muscarinic acetylcholine, Nicotinic acetylcholine	C,P
Aspartate	NMDA	C
Dopamine	Dopamine	B
Epinephrine	α_1, β-adrenergic, β_2	P
Gamma-aminobutyric acid	GABA A,B and C	C,Retina
Glutamate	glutamate (metabotropic), NMDA, Kainate, AMPA	C,P
Glycine	Glycine, NMDA	C,P
N-Acetylaspartylglutamate	mGluR3, NMDA	C,P
Neuropeptide Y	Neuropeptide Y (5 subtypes)	B
Norepinephrine	α_1 and α_2 adrenergic, et al.	C,P
Serotonin (5-HT)	5-HT	C,P
Nitric oxide	?	B

*C=Central Nervous System (B=Brain), P=Peripheral Nervous System

1.2.2   NEURAL CODING

The term *information*, in the context of our discussion, refers to the intelligence, (i.e., knowledge about something) that is being exchanged, represented, signaled, or relayed, within or between neurons; information is what is conveyed by a spike train. The coding of stimuli into trains of action potentials is complex and better understood in some systems than in others. The following discussion covers neural coding related to the topics covered in the following chapters; it only scratches the surface of neural coding theory and research. The term *Neural code* refers to a scheme or heuristic that is used by neurons to convert their inputs into trains of action potentials. In 1961, Horace Barlow postulated in his "Efficient Coding Hypothesis" that neurons use a code which minimizes the number of spikes needed to transmit a given signal [11]. Since then it has been demonstrated that single neurons can convey information with exceptionally high efficiency that approaches the theoretical limits on information transmission [12, p. 17] and it is theorized that significant information can be relayed by as few as one or two spikes. Furthermore, Barlow suggested that this code was particularly well suited for the representation of visual and auditory stimulus such as that encountered in the natural environment. This is not surprising considering that nervous systems have been optimized through millions of years of evolution in the "great outdoors" to make maximal use of their biological components.

There are a number of theories regarding how neurons or ensembles of neurons represent their information. The earliest work on neural coding was conducted by Adrian et al. [13] in 1926 who described what is known as the "rate code model." This type of coding was observed in muscle stretch receptors whereby the rate of firing of the receptors increased as the amount of weight hung from an excised muscle was increased. Thus, the frequency of the generation of action potentials was directly proportional to the stretch of the muscle. The rate code disregards the fluctuations that occur in the spacing (inter-spike interval) of spikes in a spike train. Rate codes are usually calculated over relatively long time periods which minimizes the impact of inter-spike variations.

In contrast to the rate code, the temporal coding model posits that information about the stimulus is encoded in the precise inter-spike timing of spike trains. For example, in Figure 1.8 we see a train consisting of six spikes (numbered 1–6) which occur over a 1 sec interval. If we were only to consider the rate we would say this is a 6 Hz spike train. However, we immediately notice that the intervals between the spikes are not consistent. Furthermore, there are three identical intervals of 166.7 ms and two 250 ms intervals. Using this information we could then construct a histogram of the inter-spike intervals shown at the right of Figure 1.8. The theory that information about the applied stimulus is conveyed in recurring timing motifs present in the inter-spike intervals of the spike train is referred to as correlation coding. As can be seen in the example, both rate and temporal codes can be incorporated into the same

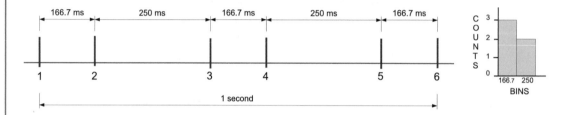

Figure 1.8: Rate and temporal coding example.

spike train since variations in the inter-spike timing can occur without changing the mean rate taken over a given time interval. This type of dual coding has been observed in the human cochlea (the part of the ear that encodes sound), the retina, and other sensory systems of the human body. Coding of auditory signals by the cochlea is of particular importance to cochlea prosthesis research (discussed in Ch. 3) which attempts to restore a type of hearing to the profoundly deaf. It should be noted that the example shown in Figure 1.8 paints an ideal picture of spike timing. In practice, there is considerable "jitter" in the inter-spike timing such that the histogram would contain bins which span a range of time values instead of the single values shown in the example.

The representation of movement direction is important for a number of systems including the vestibular[5] and motor[6] systems. The full range of all possible movement directions is represented in the activity of populations of neurons, where each individual neuron in the population responds best to movement in a specific direction. In Figure 1.9-A, we see a single neuron with a receptive field[7] whose center is 270°. The tuning curve below it indicates that while its peak response (and preferred direction) is 270°, it also responds to movement ± 15° from its center but with decreasing strength. This neuron is part of a larger population of neurons (Fig. 1.9-B), each having their own preferred direction. If we think of each neuron's preferred direction as a vector whose magnitude (length) is equal of the strength of the response, we can represent the population response as shown in Figure 1.9-C. Summation of all the vectors will produce a net population vector whose direction matches the direction of movement of the limb. Figure 1.9-D illustrates how the activity of the population and the net population vector might look for different directions of movement. The population response of neurons in the primary motor,

[5] Involved in balance and determining head position.
[6] Directs the voluntary movement of limbs.
[7] The region of physical space in which a stimulus will produce a response in a given neuron.

pre-motor, and supplementary motor cortices is under investigation by a number of laboratories as the basis for direct neural control of a prosthesis and other brain-computer interactions.

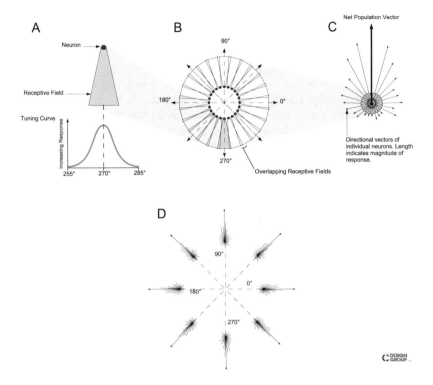

Figure 1.9: Motor cortex population coding example.

Neural coding is usually discussed from the standpoint of electrical activity vis-á-vis the spiking activity of neurons. It is important to note that the electrical activity which is so obvious and relatively easily to record, is the end product of a myriad of underlying chemical, molecular, and genetic processes which have only been discovered in the past 80 years or so. It is only in the last 40 years that we have been able to study neurochemical processes directly in living tissue and tools for the *in vivo* study of molecular/genetic events are still in the early stages of development. However, it is these underlying processes that endow the brain with its unique abilities and when things go wrong, lead to neurological deficits. To elucidate the rules which govern neural coding, we will have to dig deeper than just the electrical activity. Understanding the neural code is only the beginning, our true interest lies not with the neural code *per se* but the information that is contained within. That information, like encrypted e-mail, can't be viewed until the code is cracked. Neural coding is an important piece of the "brain puzzle" that must be deciphered in order for us to gain a better understanding of the language neural interfaces need to use when "talking" to living neurons[8].

---

[8] For further reading on neural coding consider, "Spikes: Exploring the Neural Code"[12].

1.2.3   LEARNING AND PLASTICITY

In terms of learning, two questions of importance to neural interfacing are:

1. What are the cellular mechanisms of learning? and,

2. How is memory represented in the brain?

The first question relates to determining the physical change that occurs in the neural substrate when we learn something. The volume of literature on how learning may occur in the brain is large. In general, the prevailing theory is that learning in a biological neural network manifests as changes in synaptic connectivity and efficacy. The term *synaptic efficacy* refers to the relative amount of influence that pre-synaptic activity at a given synapse has on the probability that the post-synaptic neuron will fire in response. As previously discussed, chemical synapses have the ability to modify the neural message through various mechanisms. Neurons also have the ability to add or delete (prune) synapses from their arbors. It is believed through these two mechanisms—changes in synaptic connectivity and modification of synaptic efficacy—formation of memory in the brain is accomplished. The more difficult part of the answer to the first question is not *what* changes but *why*.

The Canadian psychologist Donald Hebb postulated that:

> *"When an axon of cell A is near enough to excite cell B and repeatedly or persistently takes part in firing it, some growth process or metabolic change takes place in one or both cells such that A's efficiency, as one of the cells firing B, is increased"* [14, p. 62].

This idea is often expressed by the shorter phrase, "Neurons that fire together, wire together," and is termed *Hebbian Learning*. Hebbian learning pre-supposes the existence of a mechanism (or mechanisms) whereby the postsynaptic cell could back-propagate a signal to the active synapses which would indicate that their efficacy should increase (Fig. 1.10). In our earlier discussion, I indicated that axons conduct action potentials with no mention regarding the same for dendrites. The axon-only theory of AP conduction was the prevailing view until 1994 when research led by Bert Sakmann [15] discovered that certain neurons are able to back-propagate action potentials through their dendritic arbors. This discovery provided the basis for which Hebbian learning might occur in biological neural networks. That is, the coincidence (or near coincidence) of arrival of an AP at both the pre- and post-synaptic membranes could signal changes in the synapse whereby its efficacy would be increased. Alternately, one could speculate that lack of coincidence might, over time, result of the weakening of synaptic efficacy. Three years later, both Hebbian learning and back-propagation were demonstrated in mammalian neural networks by Markham et al. an Stuart et al. [16, 17]. However, only certain types of neurons have been shown to propagate dendritic action potentials. Since other neuron types would also have to be capable of modifying their synaptic efficacy, learning through back-propagation of action potentials is probably not the only type of learning that occurs in biological neural networks.

The second question, "How is memory represented in the brain?", may prove much harder to answer because it relates to how the synaptic modifications discussed above are used to encode either the external sensory stimulus or internal thoughts/ideas that are to be stored for later recall. Unlike the synaptic learning that occurs at the neuronal level, memory representation may not have a easy, direct physical correlate. There is a great deal of speculation regarding memory representation and cognition but little hard fact indicating how this is accomplished. Using scalp electrodes to record the EEG[9] we can see that various parts of the brain are active during different cognitive tasks. However, the spatial

---

[9]EEG: Electroencephalography. A technique which noninvasively records the brain's electric field using electrodes placed on the scalp.

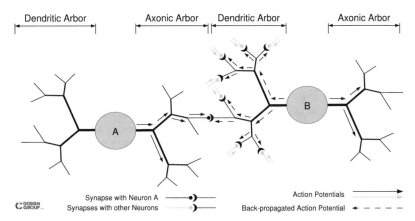

Figure 1.10: Back-propagation and Learning. An axon from neuron A synapses with a dendrite from neuron B. The sum of all the activity in the dendritic arbor of cell B, including that of other neurons that synapse with B, results in an action potential being generated a the axon hillock of B. An action potential propagates through the axonic arbor of B and back-propagates up the dendritic arbor. When the AP arrives at the terminal ends of the dendritic arbor it triggers a change in efficacy at the synapses which were active around the time of the events which triggered the AP.

resolution (pixels per square cm) of EEG is not sufficient to determine what is happening in those brain areas in great detail. Conversely, functional Magnetic resonance imaging (fMRI) has better spatial resolution but poor temporal resolution (pixels per second). Thus, smaller regions can be imaged but the image is a summation of all the activity occurring in a given time window. The current tradeoff between these and other available techniques for studying whole brain activity force the researcher to choose between either spatial or temporal resolution. What is needed is a technique that can provide both and at significantly better resolution than what is available at present. In part, the answer to the memory encoding/representation question is one which is awaiting better technology. Theories relating to memory representation will be discussed in the section on cognition.

"Plasticity" refers to the brain's ability to change and/or adapt. Synaptic plasticity discussed above, is necessary for learning while *neuro-plasticity* refers to the ability of the brain's regions to alter their function or organization. This can happen following brain damage as uninjured areas attempt to assume the damaged area's function and is usually more pronounced in younger patients. Neuroplasticity is also observed in children that are born without a particular sensory modality. For example, in children that are blind from birth there seems to be a re-tasking of some of the parts of the brain normally involved in visual processing to sub-serve other functions. However, individuals who were born sighted and blinded later in life do not exhibit as great a degree of functional re-assignment. While the brain of a child exhibits greater plasticity than that of an adult, current research [18] suggests that the adult brain may be far more plastic than was originally thought.

Both synaptic- and neuro-plasticity are important to neuroprosthesis research because the brain's ability to adapt can assist with the integration of, and communication with, artificial components. It is well known that there are critical periods during development where sensory input is required to assure the proper development of the brain. For example, children who are born with congenital cataracts must have them removed before 5 years of age or their visual perception skills will not develop normally. There is also a critical period for language development which is why cochlear implants have the highest rate of success if implanted prior to 2 years of age. In the future, children who are born blind may have the best outcome if they are fitted with a vision prosthesis early in life. This would permit the brain to adapt to the new input and integrate it as a normal part of the patient's sensorium. It is likely that this will be true for all persons who were born without a given sensory modality; sensory replacement at an early age produces the best outcome. Perhaps future biotechnology will produce a way to induce a development-like plastic state in the adult brain for the purpose of neuroprosthetic integration; until then we must rely on the brain's natural abilities. The story is different for those who were born with a modality and later lost it because their brains have already learned to process the sensory information. In this case, the sensory input must be restored before functional re-assignment begins to occur.

1.2.4   COGNITION

The term "Cognition" is an abstract concept which collectively refers to the mental processes of learning, memory, perception, judgment, and reasoning. In psychology and the neurosciences, it is usually used to refer to human (or human-like) processing and application of knowledge. Cognition is related to the idea of "mind" in that both terms refer to the conscious stream of thought. Both terms have also been applied to metaphysical constructs which posit the existence of a "soul." For the purposes of our discussion, the term "mind" will be used to refer to aspects (both known and unknown) of cognitive function which act in synergy to produce the self-aware, human mental experience.

Augmenting human cognition using artificial means is not as cut-and-dried as is often portrayed in popular literature. There are many competing theories on how the sensory inputs, biological processes, thoughts, and emotions work together to create what we refer to as our cognitive experience. Restoration/augmentation of the senses or connecting end-effectors to the motor centers of the brain may prove to be far easier than augmenting cognition. In the near future, we may discover drugs which improve memory, heighten alertness, or sharpen mental skills. Cognitive supplements are sought after by a number of government agencies and the military to improve safety for personnel who must endure long shifts with little or no sleep. Such conditions are encountered during natural disaster relief operations and on the battlefield. The more ambitious goal of building replacement parts for the brain or correcting defects of cognition such as hyper-violent behavior and schizophrenia rely on a firm understanding of neural coding and neurophysiology. The far-reaching, science fiction-like applications such as computer database integration with memory, seamless integration of digital and mental computation, or the connecting of minds together to form a powerful meta-consciousness capable of understanding far more than a single individual, will not occur without an understanding of how the physiology of the brain relates to the emergent property of mind. Ultimately, there may not be a easy way to separate physiology from cognition; the two may be inseparably intertwined. One thing is certain though, until we understand cognition it will be difficult to determine how to augment it in any practical fashion or cure cognitive defects such as schizophrenia.

Scientists often create models (computer simulations) to test their theories; most of the models of cognition discussed below involve deconstructionist methodologies (the whole is equal to the sum of its parts). Under this approach a complete description of the brain and cognitive function, relies on proper description of all of the units of which it is comprised. This begs the question, "How can we be sure that we know what all the parts should be?" Ostensibly the answer would be, "When the model produces results that match observed behavior." However, we should consider the possibility that a model might produce results that corroborate the observed behavior without accurately modeling the underlying processes that generated the behavior. That is, just because we can make a robot behave like a bug doesn't necessarily mean that we understand how or why a real bug's brain creates those behaviors. There are *many* theories regarding how the brain's physiology relates to mental function and cognition; four of them (of which I am most familiar) are discussed below.

1.2.4.1   Connectionisim

Connectionists posit that the brain's cognitive phenomena is an emergent property attributable to its network of interconnected simple units. Transformation of information occurs through the associative connections between the units comprising the network instead of symbolic manipulation. Connectionism is in opposition to the classical theory of neural processing which states that information processing in the brain occurs through the manipulation of symbols like in a digital computer. Connectionist models employ artificial neural networks (discussed in Sec. 1.3.1) to model their biological cousins. The artificial neurons can be as simple as an equation which governs how their output is related to their input. Alternately, sophisticated modeling software such as NEURON[10] or MCell[11] permits the inclusion of details such as ion channels, neurotransmitters, axonic/dendritic arborizations, and more.

Neural networks based on Connectionist models have many uses including identifying trends in data for financial forecasting, interpreting data from chemical sensing arrays for toxin detection, and other pattern recognition tasks. In terms of cognition, there have been many notable successes of these networks for simulating certain aspects of human behavior. One of first is NETtalk[12] [19] which simulates language acquisition in a small child. When the simulation begins it mostly sounds like phonemic nonsense reminiscent of the babbling sounds made by babies. The system rapidly progresses to correctly assembling the phonemes into words then connecting the words into sentences. Finally, the system is able to read text with suprising accuracy and sounds eerily like a 7-year old child, although the voice pitch, and timbre is the intentional choice of the designers.

Connectionist models have been used to model simple behaviors and have found application in a number of other areas. The successes of artificial neural networks based on Connectionists theories might lead one to conclude that this is the correct explanation of how information processing is accomplished in the brain. However, as is frequently the case in science, things are not so straightforward because other competing, and/or complementary theories have also produced positive results.

1.2.4.2   Modularity of Mind

Modularity of mind theory (MoMT) attempts to define a cognitive architecture based on the hypothesis that the brain is composed of innate, functionally distinct modules that interact to produce complex behavior. According to the theory, this architecture arose as a result of evolutionary pressures as natural selection acted to produce many of the modules [6]. MoMT proponents fall into two camps: those who believe *all* cognitive processes are modular, termed *massive modularity* and those who believe that only low level cognitive processes are modular. An example of one such cognitive module is language acquisition. All normal humans are born with the innate ability to acquire language. Damage to the language areas of the brain results in problems with linguistic expression and/or comprehension of spoken language. According to the theory, we are genetically "hardwired" with pre-defined circuitry for various modules (e.g., the language acquisition module) because of evolutionary processes.

---

[10] http://neuron.duke.edu/
[11] http://www.mcell.cnl.salk.edu/
[12] http://www.cnl.salk.edu/ParallelNetsPronounce/index.php

1.2.4.3   Subsumption Architecture

Subsumption architecture (SA) is similar to MoMT in that it too posits that complicated behavior is composed of simpler, modular elements. The difference lies in the way the modules are organized and interact to produce the observed behavior. Whereas MoMT suggests an additive and perhaps nonlinear process, SA is based on the idea that behaviors compete for dominance from moment to moment. The dominant behavior is the one that is selected by internal goals and the external environment. For example, a "search" behavior might rely on a number of simpler behaviors such as those that control movement and obstacle avoidance. When an obstacle is encountered the move forward behavior is "subsumed" by the obstacle-avoidance behavior. Once clear of the obstacle, the move forward behavior then re-assumes dominance.

Subsumption architecture has been implemented in a number of robots produced by Rodney Brook's lab at MIT. In 1990, Dr. Brooks and others from MIT founded iRobot Corporation. The company produces a number of commercially available robots including the Roomba™ vacuuming appliance. The software which guides Roomba™ is an implementation of the Subsumption architecture that mimics insect behavior. For those who like to tinker, iRobot sells developer kits, add-ons, and other accessories which permit users to hack the software and modify the hardware[13].

1.2.4.4   Holonomic Theory

The Holonomic Theory of Cognition [3, 4] is probably the most difficult to understand of all the theories discussed here. In addition to neuroscience, it borrows ideas from laser holography, gestalt psychology, and quantum mechanics. The theory was first proposed by Karl Pribram in the early 1960's based upon his studies of cortical [dendritic] receptive fields whose properties he described using Gabor[14] functions.

We have all seen holograms before, from the small squares making our credit cards more secure to novelty portraits whose eyes seem to follow us around the room, they have become a standard feature of life in the 21st century. These holograms, which are viewable in polychromatic[15] light, are called "sun-lit holograms." However, the first holograms were made with lasers[16] and lasers were needed to view them. To make this type of hologram the beam from a laser was split in two. One of the beams (termed the reference) illuminated a photosensitized glass plate directly. The other beam illuminated the object being photographed. The laser light reflected from the object is focused on the plate where it mixes with the reference beam to create an interference pattern which encodes both the scene and the viewing angles. After the plate is developed it can be re-illuminated using only the reference beam to reveal the 3D image of the object. One of the curious properties of the holographic plate is the distributed nature of the information it contains. That is, the interference pattern does not represent the corresponding points in space of the visual scene like a photograph. If the plate is broken and one of the pieces is illuminated, it does not contain a fragment of the image. Rather, it contains the whole image but viewed from an angle that corresponds to the piece's place in the original unbroken plate.

Gestalt psychology attempts to define basic principles or laws of cognition which determine the way in which we perceive objects. Gestalt theory maintains that the brain is analog, self organizing, and must be studied as a unit rather than as the sum of its parts. A gestalt is, "... a configuration, pattern, or organized field having specific properties that cannot be derived from its component parts[17]." Thus, gestalt psychology treats the mind as a unified whole comprised of gestalts which may function separately or

---

[13] http://irobot.com/sp.cfm?pageid=373
[14] Dennis Gabor (1900-1979): Nobel laureate, mathematician, and the father of laser holography.
[15] Light composed of many different wavelengths/colors (e.g., sunlight).
[16] A coherent (one phase), monochromatic (one wavelength) light source.
[17] *Websters Unabridged Dictionary*, 2nd ed., Random House Inc., New York, 2001.

together. Gestalt systems exhibit one or more of the following four key properties: emergence, reification, multi-stability, and invariance.

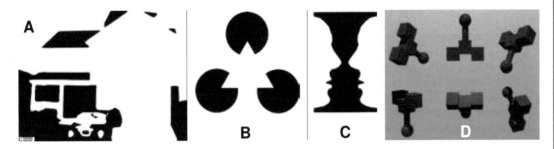

Figure 1.11: Examples of the Four primary Gestalt properties. (A) Emergence, (B) Reification, (C) Multi-stability, and (D) Invariance.

Emergence is a property that describes how our brains extract meaningful patterns when presented with incomplete information. The thresholded picture of a house with a car parked in front (Fig. 1.11-A) lacks most of the detail present in the original photograph. Yet we are easily able to identify what is depicted. This is because we have pre-existing abstract concepts (or gestalts) of "house" and "car" which our brains are able to apply to extract meaning from the incomplete/low detail patterns. *Reification*, or the property of illusory contours, can be seen in Figure 1.11-B. Although the drawing consists only of three circles with missing wedges, our mind fills in the missing contours allowing us to perceive a triangle where none actually exists. *Multi-stability* relates to perceptually unstable objects which seem to "flip" between one interpretation and another. For example, in Figure 1.11-C do you see two faces or a vase? Lastly, *invariance* describes our ability to recognize objects regardless of their rotation or scale (Fig. 1.11-D).

Simply stated, the Holonomic Theory of cognition describes a theoretical system where information is represented in gestalt form as holograph-like interference patterns. The major points of the theory are:

1. Holonomic processes are created by the electrical activity occurring in the smallest fibers of axonic and dendritic arbors in certain areas of the brain.

2. The receptive fields of these arbors are mathematically described by Gabor functions.

3. The interference patterns do not exist as "waves," instead they represent the intersection of waves and take the form of Fourier[18] coefficients.

4. All memory is not holonomic. Memory has layers, the deepest of which is the "deep holonomic store." The holonomic store is separate from the surface pattern created by the neurological storage circuitry. The neural circuitry of the brain contains the "indexing information" needed to access the deep store. For example, if you had a cube filled with thousands of different images all jumbled together, and each of the images was stored using a unique process whereby it could be retrieved using the identical process that stored it, then a great number of images might be stored in the same medium so long as each was stored using a unique encoding process.

[18]Fourier Analysis: the representation of a wave function as the sum of sine and cosine functions.

If the purpose of the neural substrate is to both support holonomic processes and permit access to the information stored within them, we must study the relationship between physiology and holonomy. Karl Pribram stated in a recent article on the holonomic theory that, "...holographic and holonomic processes are truly holistic in that they spread patterns everywhere and everywhen to entangle the parts with one another. In this domain, space, and time no longer exist and therefore neither does causality"[19]. If Dr. Pribram is correct, especially from the standpoint of the lack of causality, it could make direct, artificial access to long-term memory extremely complicated. In general, the way we learn about brain function is to observe the response elicited by a given stimulus; that is, the effect that is produced by a cause. If cause and effect break down at the level of holonomic processes then a new paradigm for their *in vivo* study must be found. It would not be enough to implant an array of electrodes in a key area of the brain, produce random stimuli and learn from the responses. In order to access the information we would have to understand how it is encoded, stored, retrieved, and decoded, then determine the exact set of neural signals needed to manipulate the underlying neural substrate to accomplish reading/writing of the deep store. Without such an understanding it would be like trying to write a computer program by randomly pressing the keys on the keyboard.

1.2.4.5   Which Theory is Correct?

I selected these four examples to give you an idea of the diversity of opinion that exists regarding cognition and the neurophysiological basis for cognitive processes. We must also consider the possibility that two or more of the above theories (or others that I did not discuss) could be correct but for different areas of the brain or for different "levels" of processing. For example, MoMT may describe phenomena like our language acquiring ability while Subsumption might be the right model for the organization of simpler behaviors, especially in the primitive brain areas and the spinal cord. Connectionism may describe the way neural circuitry organizes to process sensory data, produce modules, or create surface patterns which access the deep holonomic store, etc., etc. Based on the history of our understanding of cognition, it is likely that these theories are neither entirely correct nor entirely incorrect. Rather, like the works of Plato, Aristotle, and Descartes, they represent milestones along the road toward a better understanding of the mind and brain. No doubt future scientists will look back at our work an marvel at what we managed to get right even with our primitive techniques and have a good laugh at how wrong we were in other regards; c'est la science!

1.2.4.6   The Brain as a Mirror for itself

Ultimately, it may not be a question of whether or not we *will* understand the brain rather *can* we gain a complete understanding of brain function. Hence, the question:

>   Is there a fundamental limit to human-kind's ability to understand our own brains?

There is an enormous amount of complexity in both the way the brain is organized at the macro/micro-scopic level and what we think we know of its function. To understand the issue of the decidability of brain function you must first understand the problem space. One way of characterizing the complexity of a system is to estimate the degrees of freedom, that is, the number of ways things can change at any given instant in time. We can only make a guess at this since our understanding of the brain is far from complete and it is almost certain that we do not know all the ways things can change. Given what we currently know, the number of parameters which are freely available to change at any given moment are immense.

One way to visualize the present state of our knowledge of the brain is as a continuum that ranges from what we know with 99.99% certainty to what has yet to be discovered (Fig. 1.12-A). Among the

---

[19]Karl Pribram: http://www.scholarpedia.org/article/Holonomic_brain_theory

things we know are facts such as these: the brain contains neurons, neurons have axons and dendrites and pre-synaptic membranes secrete neurotransmitters in response to action potentials. In Figure 1.12-A, as we move to the right our knowledge is less certain. When we discuss theories that are based on this area of knowledge we tend to increase our use of qualifiers such as "possibly", "may", or "... evidence suggests that... ." As we move further to the right the probability of error in our knowledge base increases and we tend to be more abstract in our discussions. So the real question is, "How far along the continuum will we be able to progress while maintaining a high degree of certainty in our knowledge?" Unfortunately this is not a question that is easily decided, if at all.

The mathematician Kurt Gödel demonstrated via his two "Incompleteness Theorems"[20] the inherent limitations of mathematical proof. Simply put, this means that there are certain mathematical statements that cannot be decided. An example of one of these undecidable problems used by Dr. Stanislaw Ulam [21] when discussing his thoughts on the decidability of brain function uses the phenomena of twin prime numbers. Prime numbers are those integers that are divisible only by themselves and one. Some prime numbers like 3,5,7, ... , 41,43, ... differ only by two. These prime number pairs are referred to as twins. The question is: are there a finite or infinite number of twins? The fact that there are an infinite number of primes has been known since Euclid's time, however, the question of the number of twins is not known and may be undecidable since the problem space is potentially infinite. In fact, the question of decidability may itself be undecidable. This is philosophically relevent to our understanding the human brain and mind because the same question occurs: is its scope finite or infinite.

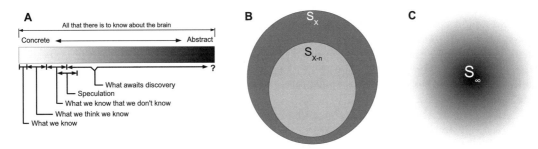

Figure 1.12: What can the brain know about itself? (A) Knowledge as a continuum from the known to the unknown, (B) The brain/mind as a finite (bounded) system, (C) The brain/mind as an unbounded, infinite system.

For example, in the case of a bounded, finite system "S" with complexity "X" shown in Figure 1.12-B the system is unable to fully describe itself because its capacity is being utilized by the very cognitive processes it seeks to describe. Thus, while the system might be self aware and able to ponder it's own existence, a full description of itself is not within its scope. However, a description of a less complex system (S_{X-n}) is within it's scope. On the other hand, if cognitive processes are not bounded (Fig. 1.12-C) as suggested by holonomic and other holistic theories of cognition, then we may be able to more closely approach a full [practical] description of the brain and all its processes. Still another possibility is that this is not the correct way to think of cognitive processes and a complete description may or may not be possible for completely different reasons than those discussed above. The problem with the "problem" is that apart from divine intervention we have no way of determining the depth of our ignorance or the height of our knowledge relative to the totality of what there is to be known about the brain/mind. At this point in time, the question of the limits to our understanding is largely a philosophical matter. I'm reasonably certain that we are not near these limits and the coming decades (perhaps centuries) will, no doubt, see remarkable progress in the area of neural interfacing and neuroengineering.

1.2.5   MACHINE INTELLIGENCE AND MODELLING

Artificial Intelligence (AI) or machine intelligence is not a field which studies human brain function per-se but is often interested in modeling cognitive processes. Such cognitive models may utilize computer simulations that are derived from, or inspired by, biological neural systems. As such, AI has a close association with neural modeling in which brain function is studied by creating computer simulations of biological neural systems. The simulations embody various properties of the system being modeled and allow researchers to study its function in detail, in real time. As computer power increases over time more complex models will become possible, however, it is doubtful that computers, as we know them today, will ever be able to simulate the entire human brain with all of its electrical, chemical, genetic, and other processes. That is, using computer models to try to understand everything about the brain is a lot like trying to use a magnifying glass to understand everything about a bacterium. The tool is simply not adequate for the job.

Dr. Hans Moravec predicted in his book, *Robot* [22, p. 63], during a discussion on the next generation of computer chips that, "As production techniques for those tiny components are perfected... The 100 million MIPS to match human brain power will then arrive in home computers by 2030." This statement is predicated on the belief that we understand or can quantify the processing capacity/bandwidth of the brain as we can with computers; this belief is quite possibly in error. While the power of computers *has* grown rapidly over the past 60 years the predictions of machines imbued with human intelligence made by various AI researchers since the 1950's have not come to pass. Don't get me wrong, some really remarkable innovation has and will continue to come from AI research. Despite 60 years of innovation, we are still not close to having an intelligent computer like the HAL-9000 of *2001: A Space Odyssey*. Not that this is a bad thing since that particular fictitious computer went insane and killed most of it's human shipmates.

All jokes aside, the reason for the languid progress of implementing human intelligence in a machine is relatively straightforward—those early predictions were made without an adequate understanding of the robustness/complexity of neural computation, in general, and of the human brain in specific. The fact that the available power of computer hardware is ever-increasing does not solve the problem of implementing human-like intelligence in software. Until we understand brain function far better than we do at present it will be very difficult if not impossible to implement that in a machine. What we are more likely to end up with are machines with a type of limited and/or highly specialized intelligence that makes them far easier to use than today's home computers. In addition to the implications for the advancement of AI, such machines will be extremely useful as the processing intermediary for neural-prostheses. Neural data processing strategies used today rely on software that interprets the detected neural activity into control directives. Smaller, faster computers, and the ability to design "intelligence" into the software and/or hardware should prove extremely useful to neuro-prosthetics.

1.3   ACQUIRING AND USING NEURAL DATA

Real-world systems are dynamic and often nonlinear. That is, they are always changing and do not respond in exactly the same way every time when given the same inputs. A single neuron, when stimulated, will not produce exactly the same response even if the stimulus is identical from one trial to the next. This seeming unreliability in the spike timings is referred to as "jitter." It is possible that what we call noise (or jitter) in spike timings is a purposeful feature of neural processing that we do not as yet fully understand [23]. It has been suggested that noise in the neural response prevents the system from falling into local minima during various types of operations. This is analogous to rolling a marble down a bumpy incline where the valleys between the bumps are the local minima. The goal is to get the marble all the way down the incline without getting stuck along the way. To prevent the marble from stopping short of its goal we add "noise" to the system by shaking the incline. The noise keeps the marble bouncing around and permits gravity to move it to its final goal.

This jitter in the spike timing means that we cannot rely on strict uniformity in the neural response. So a paradigm of, "IF you see a spike train with this timing THEN do this," will not work reliably enough to be useful. Instead, we have to utilize methods of analysis which do not depend on strict repeatability. Such methods might include those based on statistics, nonlinear dynamics, and information theory.

There are many ways of visualizing patterns in data derived from nonlinear systems, two examples are scatterplots and attractors. A scatterplot is comprised of measurements of two parameters taken repeatedly over time. If some of the instantaneous values of one parameter have associations/dependencies on the other, the data points would cluster as can be seen in Figure 1.13-A. If the values are not correlated they would scatter, forming a random pattern of dots (Fig. 1.13-B).

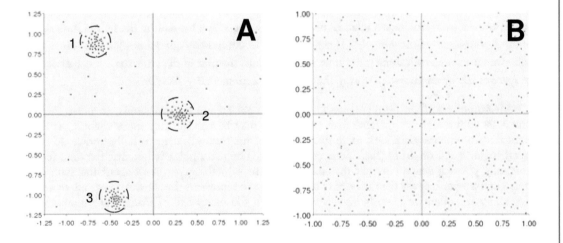

Figure 1.13: Example showing a scatter plot (A) of data which exhibits a high degree of association and (B) a plot of random (nonassociated) data. In (A), many of the recorded measurements cluster in three distinct regions of the plot. While no two measurements are exactly the same we can see both a pattern to their distribution and their bounds represented by the circles. Scatter plots are useful for showing nonlinear associations between different variables of a dynamic system.

Like scatterplots, attractors are also useful for visualizing relationships between variables in a nonlinear system. An attractor is the set of values to which the system evolves over time. Data from dynamic systems like the brain form *strange attractors*. Strange attractors are attractors which have chaotic dynamics, that is, when visualized the sets often look like noisy orbits which seem to move about one or more centers (Fig. 1.14) or "basins of attraction." The coordinate space in which the attractor exists is termed its state space and represents all the possible values that the points which comprise the attractor may take on. Each degree of freedom (a variable) is assigned an axis in the multidimensional space.

A   **B**   **C**

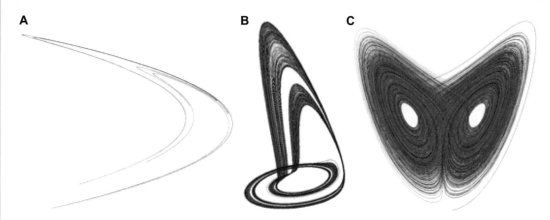

Figure 1.14: Examples of strange attractors: (A) Henon, (B) Rossler, (C) Lorenz. As the plot of the attractor evolves over time (accumulating many points), the shape formed by the points indicates the bounds on the values of the variables. If the system is in continuous operation then the values of the variables are measured over some finite time interval. Thus, the plot of the attractor is a snapshot of the state space of the dynamic system during that interval. *Generated using FractInt.*

Measurements taken from real-world systems might form attractors similar to those shown in Figure 1.14 with easily recognizable features, or they may be considerably more chaotic. Attractors constructed from neural data look more like noisy (but interestingly shaped) blobs similar to a plate of spaghetti or a roll of string that's been pulled apart. The main point is not that the data form an aesthetically pleasing shape but that the shape represent something pertinent about the state of the system. For example, let's say the the bowl of spaghetti shape represents the activity of a group of neurons during an awake, attentive state. Now, let's suppose that the messy ball of string shape represents the same group of neurons during epileptiform (seizure) activity. The state space of the attractor could be used as the input to an artificial neural network which could learn to classify the shapes into two outputs: normal and seizure. Thus, the attractor is able to represent the unique neural activity corresponding to two different states of the system.

Computers are used to acquire neural data, extract meaningful features, make decisions, and produce output for either analytical or control purposes. The initial steps in this process are illustrated in the block diagram shown in Figure 1.15. There we see a micro-electrode inserted into the brain (represented as a cylindrical volume). The electrode is connected to a headstage located a very short distance away which amplifies the detected signals. Since the signals are very small in amplitude, the headstage must be located as close to the micro-electrode as possible to minimize noise. Without some initial amplification the signals would not be detectable after traversing the longer distance to the amplifier. The amplifier boosts the strength of the signal to a range compatible with the data acquisition card installed in the computer.

The output of the amplifier is an analog waveform, that is, a signal which is continuous with respect to both time and amplitude. The DAQ card samples its inputs, periodically taking "snapshots" of the value of the function. Thus, the digitized signal is a discrete representation of a continuous function. *Aliasing* occurs when frequency components of the waveform exceed 1/2 the sampling rate. This results in an incomplete, distorted representation of the original continuous function. Most DAQ cards have built in lowpass[20] or "anti-aliasing" filters to minimize this problem.

Often amplifiers contain notch or "band-reject" filters[21] to compensate for common sources of interference such as the 50/60 cycle noise present in AC powered devices. Most amplifiers also contain at least one lowpass filter per channel. The filtering on-board the amplifier is usually intended to function as a coarse, first pass at cleaning up the signal. While the elements of the system being measured (say, two connected neurons) may exchange information with very high fidelity, noise is often introduced into the measurement, not by the cells but by our measurement technique and/or amplifiers. Once the signal has been digitized (Fig. 1.15-2), further signal processing using a computer is usually necessary to remove noise or other unwanted components (Fig. 1.15-3) and produce data that is ready to be passed to the next step.

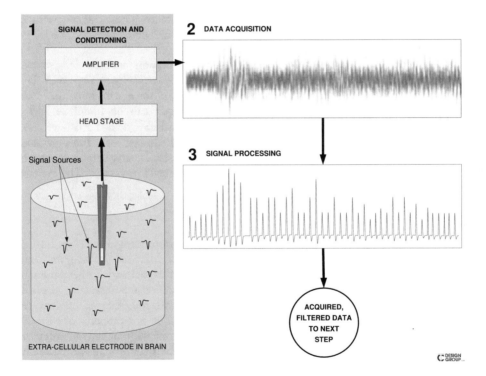

Figure 1.15: An extra-cellular electrode in the brain (shown here as a cylindrical volume) detects signals from multiple sources. The head stage is usually located very near the electrode or may even be part of the electrode. The primary function of the head stage is to amplify the waveform to a level well above the noise that will be introduced as the signal traverses the long connection to the amplifier.

[20] A filter which eliminates all frequency components above a given frequency (termed the cutoff frequency).
[21] A filter which eliminates a particular band of frequencies components.

The signal detected by the extra-cellular electrode in Figure 1.15-1 consists of spike trains from a number of sources (neurons) summed together into one composite waveform. In order to use this data to control a neuro-prosthesis with any degree of accuracy we must recover information on the activity of the individual neurons (or small ensembles) that went into the admixture. This does not necessarily mean that we need to reconstruct the exact waveform of all the sources in the mixture, although this is sometimes the goal. We do, at least, need to quantify some meaningful feature of the activity that will allow us make decisions about what to do when that particular neuron or group of neurons is active.

The general term for separating the signals from an admixture without having a-priori knowledge of the mixing process and the number of signals, is referred to as "blind source separation." The classic example of blind source separation is the "Cocktail Party Problem." In this problem there is a room filled with conversing people and microphones placed around the periphery. The goal is to separate the individual conversations from the audio recorded by each microphone. Although each microphone picks up the din of many people all speaking at once, some people are closer to some microphones than to others. Thus, the sound picked up by a given microphone would contain a different mixture of people's voices than that of a another microphone. Since the distance of a given speaker from each of the microphones varies, they are subject to delays whose phase varies as a function of distance. These differences in amplitude and phase are important features which are used to separate the individual conversations. The same is true for neural activity recorded by multiple electrodes.

Popular methods for accomplishing blind source separation include *independent component analysis* (ICA), *artificial neural networks* (ANNs), and *template matching*. ICA is a statistical technique which can be used to decompose a complex waveform into the independent sources of which it is comprised. This highly useful technique can be found in add on toolboxes for popular numerical processing software such as MatLab[22] and LabView[23]. An explanation of ICA can be found at `http://www.cis.hut.fi/projects/ica/icademo` and an online, interactive demonstration of its use in solving the cocktail party problem can be found at `http://www.cis.hut.fi/projects/ica/cocktail/cocktail_en.cgi`. The remaining two methods: ANNs and template matching are discussed below.

1.3.1   SPIKE IDENTIFICATION USING ANNS

Among the many applications for artificial neural networks are pattern recognition (Fig. 1.16) and blind source separation. The learning algorithm (rule) of the ANN dictates how the weights (connections) between elements (neurons) are modified to accomplish learning in the network. Other parameters of the ANN include the type of network (backpropagation, radial basis function, self organizing) and the number of inputs, neurons, layers, and outputs. The learning rule along with the other parameters are selected depending on the task the ANN must accomplish. *Self organization* is a property of certain dynamic systems whereby the complexity of the system may increase over time without external influence. When applied to neural networks it is used to indicate a neural network where some or all of the aforementioned parameters are determined by the network itself as it operates. Self-organizing ANNs are guided by unsupervised learning rules. One such learning rule was conceived by Linsker et al. [24] further developed by Bell et al. [25] and is based on principles taken from the field of information theory.

To begin to understand how blind source separation is accomplished using ANNs, you need a basic familiarity with *information theory*. Information theory began as a branch of mathematics used in the communications industry for studying the limits on data compression. Today, it is a broad discipline

---

[22]MatLab is a trademark of The Mathworks.
[23]LabView is a trademark of National Instruments.

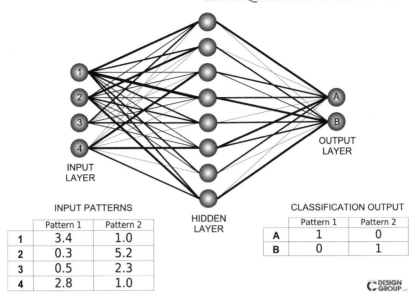

INPUT PATTERNS

	Pattern 1	Pattern 2
1	3.4	1.0
2	0.3	5.2
3	0.5	2.3
4	2.8	1.0

CLASSIFICATION OUTPUT

	Pattern 1	Pattern 2
A	1	0
B	0	1

Figure 1.16: Illustration showing the parts of a three-layer artificial neural network or "Multi-layer Perceptron" and its use for pattern recognition. In this example the network has four input neurons each of which connect to eight neurons in the hidden layer then two neurons in the output layer. The thickness of the lines connecting the layers indicates the relative importance (weight) that the receiving neuron places on the input from the originating neuron. The weights are determined during training of the neural network. The task of the network in this example is to recognize and classify a variety of patterns (two are shown) as belonging to either group "A" or group "B". The network's "decision" is indicated by a "1" at one of the output neurons.

that has found use in many other fields including neuroscience where it has been applied to the study of neural communication. One of the key information theoretic measures is the *Shannon entropy* [26] which is a measure of the complexity of a given signal. The greater the complexity of a signal the more bits that are needed to represent it across a communications channel. For example, a coin toss has an entropy of 1 bit since a single bit is all that is needed to represent the two possible outcomes (0=heads or 1=tails). If noise is added to the coin, say the coin lands on a bumpy surface where some outcomes are not clearly heads or tails, then more bits are required to accurately represent all the possible states of the coin. In other words, noise increases signal complexity thus increasing entropy.

In terms of our discussion on neural activity, the communication channel is the electrode and the signal is the waveform consisting of the summated activity of an unknown number of sources. When this detected activity consists mostly of noise the distribution of the signal is broad and has a high entropy (Fig. 1.17-A). Conversely, when the signal consists mostly of high amplitude spikes its distribution is narrow and has a lower value for the entropy (Fig. 1.17-B). The distribution of a signal that contains little noise and only a single source would be even narrower than that shown in Figure 1.17-B. However, noise is not the only source of complexity, in the brain where there are many active sources contributing to the summated waveform, the activity itself can be very complex. In this case, much of the low amplitude activity (activity that is more distant from the electrode) cannot be distinguished from noise and therefore cannot be separated.

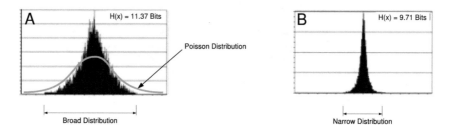

Figure 1.17: Comparison of the entropy H(x) and distribution of two signals. The distribution at the left is of a poor quality signal that contains mostly noise. As the neural data becomes more noisy its distribution approximates the shape of the Poisson distribution. The signal on the right is of a "clean" recording containing little noise. The signals were obtained from recordings of the activity in the lateral eye optic nerve of the Horseshoe crab *Limulus polyphemus*. *Adapted from [27].*

As in the cocktail party problem discussed earlier, electrodes placed in different locations around a group of neurons in the brain will "see" the summated electrical activity originating from all of the nearby sources. However, since some sources are closer to certain electrodes than others, the mixture will not be identical for each electrode. When there are two or more electrodes recording in proximity to one another, the entropies of pairs of recorded signals can be used to compute the Mutual Information or *I(X;Y)* which provides a measure of the similarity between the distributions of two signals. In this case, the signals are different "views" of the same signal-scape as seen from the points of view of different electrodes. Using Figure 1.17 as an example, the I(X;Y) between distributions "A" and "B" would be small, while the I(X;Y) between the distribution of a noiseless signal that contains only a single source and that of "A" would be even smaller. The mutual information provides us with a measure of the redundancy between two signals. InfoMax incorporates information theoretic measures into the unsupervised learning algorithms of a self-organizing ANN to extract the independent components of the waveforms using redundancy reduction. The inputs to the network are waveform data collected from multiple electrodes. The ANN then iterates with this data and learns the proper network configuration to extract the nonredundant components. The outputs of the ANN are waveforms which contain little or no redundant information. Depending on the complexity of the initial waveforms (number of summated sources + noise) the number of recoverable source waveforms will vary. At least one of the output waveforms will contain the residual components of signals which were inseparable.

One of the first of these ANN-based, blind source separation techniques was "InfoMax" which was successfully demonstrated by Bell and Sejnowski at the Salk Institute in 1995. Subsequent efforts by a number of researchers [28, 29, 30, 31] have demonstrated blind source separation using other types of neural networks and learning algorithms. The principle of redundancy reduction used in INFOMAX is reminiscent of that proposed by Barlow [11] with respect to sensory systems, in general, and by Atick et al. to retinal processing in specific. As such, this method of source identification/separation may be more "brain-like" than other methods. Since the network must re-organize (re-train) with each new data

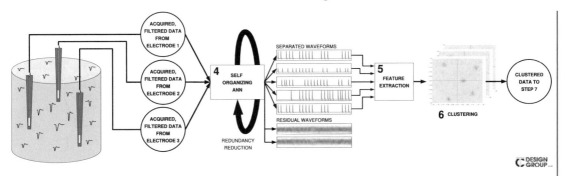

Figure 1.18: Block diagram of the technique of blind source separation by redundancy reduction using a self organizing network. The waveform data is passed to the initialized network which then interates with the data producing separated and residual waveforms. The separated waveforms are passed to step 5 while the residuals are discarded.

set, it is too slow to be of use in an overall scheme for real-time control of a prosthesis. I mention it here because it is the most robust of the three techniques and the speed issues are largely technological in nature and will likely be solved given further advances in computer hardware. The most widely used strategy at present for accomplishing spike sorting in real time is template matching.

1.3.2   SPIKE SORTING USING TEMPLATE MATCHING

In the previous section we discussed the process of de-constructing the recorded waveform using artificial neural networks. This process assumed that we knew nothing about the nature of the sources. Template matching, on the other hand, takes advantage of the fact that we *do* know something of the source; we know it is a spike train. As such, we know the general shape of the spikes which comprise the train and how their shape may be distorted by the C-R[24] properties of the extracellular medium. Additionally, signals originating from sources close to the electrode will be greater in amplitude than those from more distant sources. With this knowledge we can derive a large library of templates which allow the signal shape to be used as the basis for assigning detected activity to individual neurons.

Spikes are detected using amplitude thresholding, that is any signal above a pre-determined voltage level is considered a spike, anything below is noise. Then, using the template library the algorithm searches through the identified spikes trying to match features of the spike to the templates. When a match is found that section of the waveform is marked and the process continues. The marked segments are then overlaid as shown in Figure 1.19-4. The operator then uses a box drawing tool to indicate which group of spikes seems to have originated from the same (or similar) sources. The marked segments are then re-assembled into spike trains which contain spikes from only a single source. The individual spike trains are then processed further (Fig. 1.19-5) to extract relevant features such as timing characteristics, amplitude, etc. Those features might then used to plot the data on a Cartesian coordinate system where, hopefully, the data will cluster (Fig. 1.19-6).

---

[24]C-R: Capacitive-Resistive. In essence, the extracellular medium acts as a lowpass filter whose cutoff frequency is inversely related to the distance from the detector.

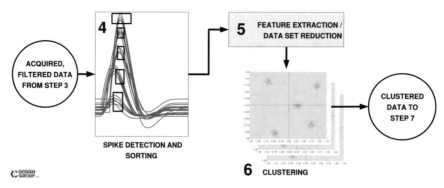

Figure 1.19: Diagram of the technique of source separation using spike sorting. Steps 1-3 are shown in Figure 1.15.

1.3.3 USING NEURAL DATA

The output of step 6 (Fig. 1.18-6 or 1.19-6) in this example are arrays of (x,y) values. Each array (or dataset) corresponds to relevent features of the detected neural activity during the time period of interest. These values form well defined clusters on scatterplots of each dataset and represent the unique state of the system (the brain region being monitored) when, for example, the subject was asked to perform a certain task. Using these datasets we can train an artificial neural network to recognize the pattern of clustering that is unique to each state. The ANN will map these states to binary output values which correspond to commands to operate a computer mouse. Thus, when one pattern of activity is recognized the move-up output might be active, another pattern of activity might correspond to move-down and so forth. The output of the neural net might then be translated into standard Human Interface Device (HID) commands and converted to the Universal Serial Bus (USB) interface protocol which would allow it to be connected to a computer and substitute for the mouse.

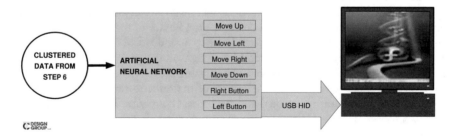

Figure 1.20: In the final step of the example the clustered data is used to train an artificial neural network (ANN). The network learns to classify the pattern of activity in the clustered data into six outputs which correspond to mouse controls and movement. The system running the ANN then outputs these commands over a USB interface to a PC to allow the user to move the cursor and interact with the desktop.

In this example, the neural activity is used to control a cursor on a computer screen. However, the neural net might have been trained to produce output suitable to control a prosthetic arm, the controls of a wheelchair or some other application. While ANNs are not the only method for performing pattern recognition they do have the unparalleled advantage of adaptability.

CHAPTER 2

Neural Interfaces: Past and Present

Through evolutionary processes biological organisms have come to make maximal use of the building blocks of their biological substrate. At the genetic level, for example, homeobox sequences that guide body patterning in early development are highly conserved (relatively unchanged) across seemingly unrelated species. In the nervous system, certain amino acids, the building blocks of protein found throughout the body, also function as neurotransmitters. In essence, biological organisms are the living embodiment of "Kaizen"[32], the principle of incremental positive change toward a goal of optimization of a system or process. Multi-use molecules are but one feature of this evolutionary optimization.

The lessons learned from the study of biological systems have diverse application in many areas including neural interfacing. The field which studies and applies these lessons is *cybernetics*. The inception of cybernetics as a discipline occurred in the early 1940's with the work of Norbert Weiner [33]. It includes areas of communication and control theory concerned with the comparative study of biological and man-made automatic control systems. The term is now also used to refer to a field which seeks to apply the knowledge gained by this comparative study to build better machines or to build equipment which in some way mimics human function. The term "Cyborg" (CYBernetic ORGanism) was coined by Clynes and Kline in their 1960 article in *Astronautics* to refer to a human connected to a machine for some adaptive purpose such as functioning in outer space [34]. In pop-culture the term Cyborg has been applied to not only the use of technology to restore function in the physically disabled but to make those restored individuals, "better, faster, and stronger" than before.

The term *Neural Interface* refers specifically to a category of technologies which are concerned with connecting excitable tissue (nerves, brain cells, muscles, etc.) to machines and is part of the broader field of Neural Engineering which is, "... an emerging interdisciplinary research area that brings to bear neuroscience and engineering methods to analyze neurological function as well as to design solutions to problems associated with neurological limitations and dysfunction"[35]. The field includes (but is not limited to) disciplines which study neural systems such as cybernetics and neuroscience, and aspects of other disciplines such as biomedical engineering, mathematics, prosthetics, and modeling as they apply to the definition of the field given above. The highly interdisciplinary nature of neural- interfacing/engineering stems from the need to pull together a diverse array of technologies from sometimes seemingly unrelated disciplines to accomplish its goals.

Human augmentation is not just on the minds of science fiction writers. The Defense Advanced Research Project Administration (DARPA) in the U.S., known for funding the more "far-reaching" research efforts, issued a solicitation on "Augmented Cognition" in 2001 which advocated research using a number of different technologies for the explicit purpose of tapping into and augmenting human sensory and cognitive function. In 2002, Lt. Cmdr. Dylan Schmorrow, Ph.D. (DARPA) said in a speech at DarpaTech that, "We are entering an era of unprecedented human advancement in which Darwinian principles of evolution may begin to show signs of artificial self-acceleration"[36] referring to the potential application of technological, biological, and psychological research to precipitate the emergence of the next stage in human development. While we are many years away from even remotely achieving Dr. Schmorrow's (and others) vision of artificially accelerating the evolution of the species, it is an intriguing idea which feeds the sometimes heated debate on the ethics of neural engineering. The present Augmented Cognition program capitalizes on the billions of dollars that have been spent by various government agencies (including DARPA) over the past 20 years, funding various segments of this work.

The National Institutes of Health (NIH), National Science Foundation (NSF), and other agencies in the U.S. have a myriad of research projects which either directly or indirectly advance neural interfacing research. Outside the U.S., a number of universities have their own independent projects underway and there are also a number of fledgling commercial efforts both in the U.S. and abroad. The primary focus of much of the work is rehabilitation and recovery of function following injury/disease or from birth defects. However, an interface developed for a prosthetic arm or a cochlear (hearing) implant might also be used as a means of directly interacting with a computer for expansion of function rather than just restoration.

Researchers have been electrically probing and recording from the brain for over 100 years, hoping to unlock its secrets. The first brain-to-machine interfaces were developed in the early 1960's and implanted in the first human volunteer in 1968 [37, 38]. However, it wasn't until the advent of micro-fabrication technology in the 1970's that it became possible to put large numbers of very small transducers into brain tissue. Since that time such devices have been decreasing in size, increasing in complexity and have enabled the study of networks of neurons *in vivo*. Some of the first individuals to experiment with brain-to-machine interfaces and neuro-prosthetics were Dr. José M.R. Delgado, Dr. Giles S. Brindley, and Dr. Walpole S. Lewin.

In the 1930's, Walter Hess[1] demonstrated that motor and emotional responses could be produced by electrical stimulation of the brain of cats. In the 1950's, building on this earlier work, Dr. José Delgado refined the procedure for the implantation of very small diameter wire electrodes into brain tissue using a stereotaxic (three-dimensional positioning) instrument and aseptic (sterile) procedures. He then used these intra-cerebral electrode assemblies, sometimes containing upwards of 100 wires, for the study of the electrical activity in various deep-brain regions. Coupled to wireless recording and stimulation devices he invented (termed stimocievers), Delgado was able to study the cerebral mechanisms of behavior in awake, freely moving animals, and humans.

The ability to both record and stimulate offered a number opportunities for experimentation including the possibility of automated learning[2] using a reciprocal brain-to-computer-to-brain connection. Demonstration of this concept came in the late 1960's when Delgado, working with other researchers, connected a female chimpanzee named Paddy (Fig. 2.1) to an analog computer. Paddy had electrodes implanted in her left and right amygdaloid nuclei[3] and her Reticular formation[4]. Using data telemetered via a stimociever the computer was able to detect the occurrence of a specific type of neural waveform (a spindle) in her amygdala then stimulate her reticular formation in response. After two hours on this paradigm the occurrence of spindles in her amygdala was reduced by half. After 6 days with treatments lasting 2 hours each day the occurrence of spindle waveforms was reduced to less than 1 percent of the untreated value. Paddy was quieter, less attentive, and less motivated during behavioral testing but still performed her given tasks without error. When the treatments were discontinued, she returned to normal after about 2 weeks. These experiments provided strong evidence that behavioral modification through automated learning was indeed possible. He predicted in his 1969 book, *Physical Control of the Mind* [39, p. 91], that this new technology could "...provide the essential link from man to computer to man, with a reciprocal feedback between neurons and instruments..."

---

[1]Nobel Prize in Medicine, 1949.
[2]Learning which occurs artificially, without the subjects participation.
[3]A part of the brain involved in the processing and memory of emotional responses to stimuli.
[4]A part of the brain involved in alertness, fatigue, and motivation.

Figure 2.1: Paddy (left) and Carlos (right) outfitted with early versions of stimocievers. Later, animals were implanted with versions of this technology which was miniaturized and contained completely under the skin. These miniature units would later be used in humans [39]. *Photo courtesy of J.M.R. Delgado.*

In addition to automated learning which had potential application in the detection and prevention of epileptic seizures, another of the goals was to identify areas of the brain involved in aggressive or anti-social behavior. Dr. Delgado and his collaborators performed extensive experiments in animals and humans throughout the 1960's to determine if electrical stimulation could be used to modify behavior. His promising results were reported in a number of papers and in the previously mentioned book. At a time when lobotomies were commonly used for treating severe forms of mental illness, Delgado's work held the promise of providing a safer and more effective alternative.

In 1966, at a ranch outside of Cordoba Spain Dr. Delgado and his team implanted wireless, brain stimulators (Fig. 2.2-A) into several adult bulls. The stimulation electrodes were implanted in the region of the caudate nucleus which is one of the structures found in the mid-brain. In a dramatic demonstration of the potential of technology to modify behavior Delgado halted a Brave[5] bull in mid-charge using nothing more than electrical stimulation. After several stimulations there was a lasting inhibition of aggressive behavior (Fig. 2.2-B) [39].

Dr. Delgado's work stirred up a furor of controversy in the 60's and was largely misrepresented by the media of the day. His work was the subject of articles and editorials in a number of publications including the New York Times magazine. Conspiracy theorists believed his work was evidence of government plots to control its population with covertly implanted devices. Conservative ethicists railed against the potential use of such technology to forcibly modify peoples behavior. The truth was (and still is) that our knowledge of the brain is far too primitive to allow this type of manipulation. However, deep brain stimulation has helped many sufferers of Parkinson's disease, epilepsy, chronic pain, and other illnesses.

---

[5]"Brave" refers to the breed of bulls used for bull fighting. Brave bulls are dangerous animals that have been bred for their aggressive tendencies over numerous generations.

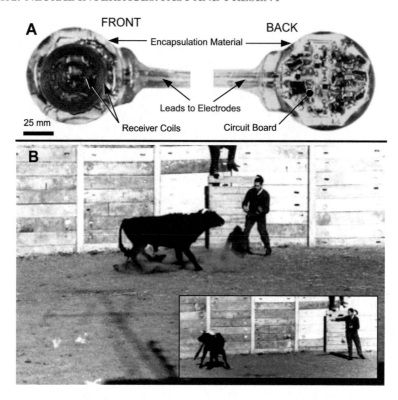

Figure 2.2: (A) Example of a four-channel wireless, deep-brain stimulator developed by Dr. Delgado. (B) Bull halted in mid-charge by inhibition of aggressive behavior using brain stimulation. Dr. Delgado can be seen holding the remote control which actuated the implanted stimulator. (B Inset) The bull not only stops but turns away from the object of his rage (the investigator) [39]. *Photos courtesy of J.M.R. Delgado.*

Delgado, now 93 and living in the U.S., says that the purpose of his work was always, "...to understand how the brain works and understand which areas of the brain were involved in aggressive behavior"[40]. He has devoted a large part of his career to understanding the psychophysiological roots of anti-social behavior. He originated a number of techniques and devices including deep brain stimulation electrodes, radio-controlled stimulators, brain pacemakers, and brain-to-computer-to-brain communication.

Around the same time as Dr. Delgado's work in the U.S., Drs. Giles Brindley and Walpole Lewin were experimenting with artificial vision in the U.K. Their work to create a vision prosthesis for implantation on the surface of the brain [37] built upon the earlier work of Foerster [41] and Krause et al. [42], both of whom experimented with electrically evoked visual sensations by direct brain stimulation. The device built by Brindley and Lewin (Fig. 2.3) consisted of an array of 81 small coils with platinum electrodes. The entire array, with the exception of the working surfaces of the electrodes, was encapsulated in a sheet of silicone rubber that allowed the array to flex and conform to the surface of the brain. The entire device was contained within the skull and had no external connections. The coils could be individually activated by placing a transmitter coil over the desired location and emitting a radio frequency pulse. The receiver converted this pulse into an electric current which stimulated the brain region directly under the electrode.

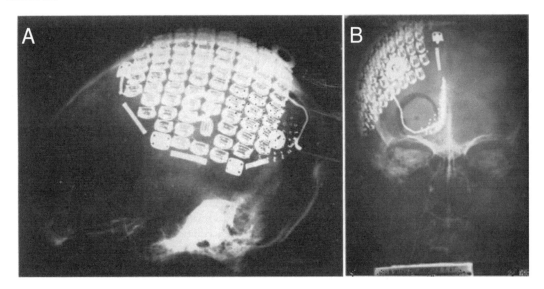

Figure 2.3: X-Ray of implanted electrode sheet. (A) Left side view. (B) Back view [37]. *Photos used with permission of Blackwell Publishing.*

When a coil in the array was activated the patient experienced the sensation of light. The sensation took the form of a small spot or spots of light (termed "phosphenes") in a constant position in her left visual field. The location of the phosphenes in the visual field corresponded to cortical visual maps already established by Holmes in 1945 [43]. The phosphenes track with the eyes during voluntary eye movement but not during reflex eye movements. Brindley and Lewin demonstrated that through activation of multiple coils simultaneously the patient could be made to see simple shapes and patterns. Later work by Brindley focused on characterizing the cortical response to stimulation [44] and its sensory effects [45]. Dr. Brindley obtained a patent for the device in 1972 [46] but his work on this vision prostheses did not continue much past 1973. However, work continued on the development of cortical vision prosthetics by others [47, 48, 49] and with the advent of microchip fabrication technologies, researchers developed high-density electrode arrays [50] for implantation in the brain. In January 2000, researchers at the Dobelle Institute in Lisbon, Portugal announced an implanted vision system which is able to restore rudimentary vision to totally blind individuals [49].

William H. Dobelle, MD devoted much of his career to cortical prosthesis research. He founded the Dobelle Institute in Lisbon, Portugal where he developed a vision prosthesis that was implanted in about six patients with good results. The system permits the user to read two inch tall letters at a distance of 5 feet through a narrow visual tunnel with acuity roughly equivalent to 20/400 vision. One of those patients, a Canadian man named Jens who had been totally blind for nearly 20 years, was able to drive a car around a parking lot using the prosthesis. With the passing of Dr. Dobelle in 2004, 30 years of work developing the "Dobelle Artificial Vision System" came to a halt. There are a few groups who continue to experiment with cortical vision implants [51] but no system like Dobelle's is presently available.

Over the intervening 40 or so years since the pioneering work of Drs. Delgado, Brindley, Lewin, and others, research has produced many technologies which are likely candidates for directly linking biological organisms to computers. The creation of a bi-directional neural interface relies not on a single technology but devices and expertise from a diverse mix of disciplines. Research efforts are necessarily multifaceted consisting of groups conducting bio-materials, interfacing, neural coding, and other research. While some groups are developing specific applications (e.g., cochlear implants to restore hearing) others are accomplishing the basic research upon which future applications will be built. Interfacing technologies can be grouped into two broad categories: those which aim to connect the Central Nervous System (CNS) and those that connect the Peripheral Nervous System (PNS) to a computer or other hardware. However, some technologies have applications in both areas.

2.1 INVASIVE INTERFACES

The term "invasive interfaces" is used here to refer to any technology which requires breaking the skin to accomplish implantation. The ability to place a sensor directly into the tissue you wish to monitor allows far more accuracy than can be obtained with noninvasive techniques. The disadvantage is the risk of sequelae such as infection, tissue reactions, etc. Furthermore, invasive devices for use in humans have a strict regulatory process that must be navigated, surgical procedures have to be performed by trained surgeons under aseptic conditions, anesthesia, and post-operative support are needed, all of which make this type of interface far more complicated to commercialize. From a signal processing standpoint, it is far less complicated and far more feasible derive useable control signals from an electrode placed in the brain as opposed to an electrode placed on the scalp. One of the factors that currently limits the useability of noninvasive interfaces is the difficulty in recovering usable control signals from the highly complex electrical activity recorded by surface detectors.

2.1.1 PERIPHERAL NERVOUS SYSTEM INTERFACES

The peripheral nervous system includes nerves which connect the senses, skin, muscles, and internal organs to the central nervous system (CNS). The CNS consists of the brain and spinal cord. A peripheral nerve contains multiple nerve fibers and is, in concept, like the trunk cable of an old style telephone system containing thousands of wires each carrying a different conversation. However, unlike the telephone cable each fiber in a nerve carries a unidirectional signal. Motor fibers carry signals from the brain to the muscles while sensory fibers communicate information on temperature, pain, pressure, vision, etc., to the brain. Some nerves contain all one kind of fiber (e.g., the optic nerve contains only sensory fibers) but many peripheral nerves are mixed. A mixed nerve complicates processing because in order to control a prosthesis we need to separate the signals originating from motor fibers from those of the sensory fibers. The nerve fibers are arranged in bundles (fascicles) which are contained within the nerve sheath (Fig. 2.4).

2.1.1.1 Cuff Electrodes

The general design for a cuff electrode is that of a flexible material containing wires on the inner surface which is then wrapped around a nerve. The edges of the material may be glued, sewn, or otherwise connected so as to form a tube and hold the bare wires in contact with the surface of the nerve. Cuff electrodes have also been fabricated from short lengths of silicone elastomer tubing (Fig. 2.5) as well as other materials. Cuff electrodes have the advantage of being relatively easy to fabricate from commonly available materials and are well tolerated for both acute and long-term experiments [52]. A self-sizing

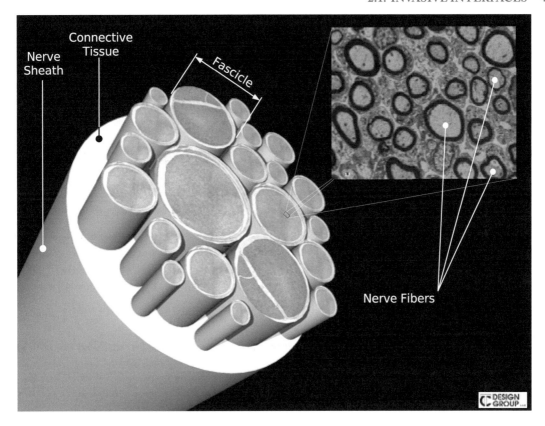

Figure 2.4: Diagram of the structure of a peripheral nerve. Nerve fibers are arranged in bundles (fascicles) which are stabilized by connective tissue and contained within the nerve sheath. Also contained within the connective tissue are blood vessels (not depicted).

cuff electrode [53, 54] can be fabricated from tubing by making a longitudinal cut which spirals from end to end. The tube is then un-coiled, wires glued into place following the spiral and the tube allowed to re-coil. When wrapped around a nerve, the cuff will expand as needed to fit. Spiral nerve cuffs have been used experimentally in humans with degenerative retinal disease as a means to directly stimulate the optic nerve [55].

2.1.1.2 Nerve Re-Shaping Electrodes
When a cuff electrode is placed around a nerve, each electrode detects the nearby electrical activity and, to a lesser extent, that occurring elsewhere in the nerve. This complicates the process of sorting out which fibers are active in the nerve at a given time. One way to address the problem of signal separation in a round nerve is to try to reconfigure the nerve to a more desirable shape [56]. That is, if the nerve were flat instead of round then the nerve fascicles would be arranged serially instead of bundled together. Dr. Dominique Durand of Case Western Reserve University and others have developed a type of cuff electrode that is

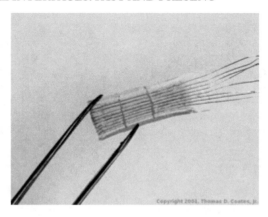

Figure 2.5: Photomicrograph of a cuff electrode [27]. The cuff is formed from a short length of silicone tubing which has wires affixed to the interior. The tube is slit parallel to the long axis to permit it to be opened and placed around a peripheral nerve. Here, the cuff is held open and flattened (using a pair of small tweezers) to reveal the wires running parallel to the length of the cuff. The cuff shown is approximately 1.5 cm in length.

capable of re-shaping the nerve. The flat interface nerve electrode (FINE) [57] is fabricated from silicone sheeting, iridium wire, and PLGA, a biocompatible polymer that slowly dissolves *in vivo*. The silicone sheeting is used to construct a flattened tube which represents the final state of the cuff. The tube is stretched into a round configuration and PLGA is applied. When the PLGA is cured, it forms a strong, rigid band around the flexible tubing which maintains it in the round configuration. Once placed around a nerve *in vivo*, the PLGA begins to dissolve over approximately a 1-month period. As the PLGA dissolves, the tube gradually resumes its original configuration and the nerve is safely flattened. Once flattened, the electrodes are able to detect activity originating from nearby sources with far better isolation than before. The FINE can also be used to stimulate nerves with better specificity than cuff electrodes [57, 58].

2.1.1.3 Nerve Regeneration Arrays

Regenerative Nerve Interfaces (RNIs), also known as sieve electrodes, capitalize on the limited ability of peripheral nerves to regenerate following injury. When a peripheral nerve is cut, the nerve fibers inside the fascicles will degenerate back to the cell bodies in the spinal cord (termed Wallerian degeneration). The cell body will then send out a new process which grows down the nerve fascicle inside the sheath toward the cut end. If the cut nerve is reconnected with the fascicles aligned many of the fibers will find their way back to their original connection sites in the limb. If a porous material is placed between the stumps, the nerve fibers will grow through it and into the distal stump (Fig. 2.6). The basic design of an a RNI consists of a thin, porous substrate (the sieve) with electrodes around some of the pores. The RNI is placed between the stumps of a cut nerve and the nerve allowed to regrow through the fenestrated substrate. The electrodes can then be used to record from and stimulate nearby nerve fibers. If sufficient electrode density could be achieved RNIs could provide fine control and feedback between a nerve and a neuroprosthesis.

Advanced RNI implementations have used silicon microchips containing perforating via holes, micro-fabricated electrodes, and on-chip amplifier/multiplexing circuitry. The early work in this field was

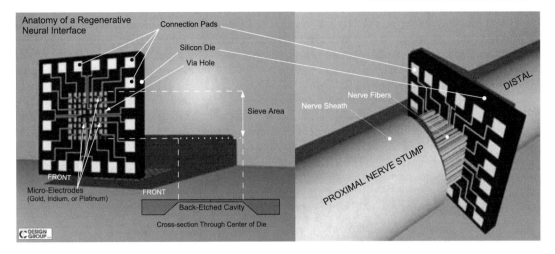

Figure 2.6: Illustration of a Regenerative Neural Interface. Left: An RNI similar to those developed in the U.S. and Europe. Right: Diagram of the RNI *in situ* (not to scale).

conducted by A. F. Marks [59] (1969) and R. Llinas [60] (1973) while trying to improve peripheral nerve regeneration following injury. Later, researchers at Stanford [61, 62], University of Michigan [63], and University of Stuttgart worked on advanced designs. Work on this type of interface still continues at The Fraunhofer Institute for Biomedical Engineering [64], University of Freiburg, and other sites in Europe and at the National Cardiovascular Research Institute in Osaka, Japan [65]. Continued progress on these types of interfaces is contingent upon advances in human nerve regeneration, biomaterials research, and micromachining techniques. Since this interfacing strategy requires nerve transection it is not likely that it would be useful for connecting devices to nerves in healthy, uninjured individuals. However, in cases of amputation (where the nerve is already cut) it could become practical in the relatively near term as an interfacing technique for prosthetic limbs.

2.1.2 CENTRAL NERVOUS SYSTEM

From the perspective of evolution, the primary purpose of the nervous systems of all animals is to allow the rapid response and adaptation to ever changing situations in their environments to increase the animal's chances of survival. The central nervous system, as already mentioned, consists of the brain and spinal cord. The spinal cord and lower brain are responsible for the reflexive responses, autonomic processes (like breathing), and lower-level processing functions. The higher brain, composed of the cerebral hemispheres and other structures, is responsible for high-level functions which gives rise to our cognitive processes and allows us to carry out voluntary actions. Interfaces designed for use in the CNS typically use arrays of penetrating microelectrodes, that is, microelectrodes that are inserted into the tissue. Microelectrode arrays can be constructed in a number of ways: using silicon micromaching techniques, using individually inserted wires, or by patterning conductive materials on other rigid or semi-rigid substrates which can then be implanted. One of the first micromachined-silicon, multi-electrode penetrating arrays was developed at the University of Utah.

2.1.2.1 "Hair Brush" Electrode Arrays

The Center for Neural Interfaces (CNI) was founded in 1995 by Richard Normann of the University of Utah. Among its goals are to develop multichannel neural interfaces, test the safety/efficacy of those devices, and develop new therapeutic approaches to treating nervous system disorders. As part of this mission new types of microelectrode arrays (Fig. 2.7) were constructed using three-dimensional microfabrication techniques [66, 67].

The present 3D microelectrode array design is the result of years of research by scientists at the University of Utah, Brown University, MIT, and Emory University. These high-density arrays have been developed to interface superficial cortical areas of the brain [68] and peripheral nerves to computers. They are used by several of the neuroprosthetic systems discussed later.

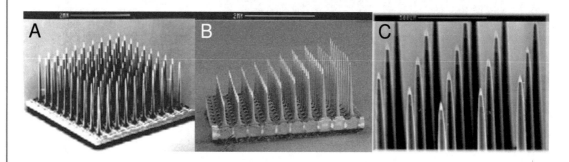

Figure 2.7: Photos of the (A) Utah Electrode Array, (B) Utah Slant Array, and (C) closeup of the microelectrode tips. *Photos courtesy of R.A. Normann.*

The Utah Electrode Array (UEA) is fabricated from silicon and contains an array of 100 penetrating microelectrodes (Fig. 2.7-A). The UEA is designed to be implanted in the cerebral cortex and is capable of stimulating and/or recording from large number of neurons simultaneously. The second type of array fabricated by Dr. Normann's team was the Utah Slant Array (USA) which is similar to the UEA except that it contains 100 penetrating microelectrodes of various lengths (typically from 0.5–1.5 mm in length). The USA is designed to be implanted in the peripheral nervous system and to provide a means for simultaneous stimulation and recording from a large number of nerve fibers.

In 2002, Dr. Kevin Warwick, a cybernetics professor at the University of Reading, U.K.[6], was implanted with an array similar to the UEA. The 100 electrode array was implanted into the median nerve of his left hand. Signals derived from the interface permitted him to control an electric wheelchair and a robotic arm. Warwick's use of himself as a research subject was, and still is, highly controversial. To date, the implants are functioning properly with no ill effects to the professor.

The UEA and USA have been used extensively in animals [69, 70, 71] and more recently in humans in conjunction with the BrainGate System®. The BrainGate® System[7] developed by Dr. John Donoghue's team at Brown University and marketed by Cyberkinetics is currently beginning clinical trials for individuals with severe motor impairment. If successful, the system will permit paralyzed individuals to interact with their environment through novel, computer-based communication interfaces.

---

[6]http://www.kevinwarwick.org
[7]http://www.cyberkineticsinc.com

2.1.2.2   Microwires

The use of fine wires for recording and electrical stimulation of the brain dates back to Walter Hess in the 1930's. Since then a number of researchers including José Delgado [72], C.A. Palmer [73], Westby [74], Richard Andersen, Joel Burdick [75], and others have refined the technique. Microwires can be placed at depth in the brain using stereotaxic equipment and a guide cannula or they can be coated with glass to stiffen them. Arrays containing hundreds of microwires can be constructed allowing detailed study of large areas of the brain.

One of the problems encountered with any chronically implanted microelectrode is that the electrode, neuron, or both can shift position over time resulting in loss of signal strength. Drs. Richard Andersen, Joel Burdick, and Yu-Chong Tai at Caltech are developing a micro-positioning unit using electrolysis-based diaphragm actuators [76] capable of repositioning the individual electrodes in a multiwire array. This MEMS device would enable the array to "intelligently" track the signal sources and reposition the electrodes as needed to maintain optimal contact. A larger version of the device [75] using piezo-electric positioners has already been constructed and the control algorithm tested *in vivo* with excellent results.

2.1.2.3   Silicon/Polymer Microelectrode Arrays

Silicon has a number of advantages over other materials for use in the fabrication of neural interface devices. Its long history of use in the electronics industry provides an extensive knowledge base of how to fabricate microelectronics in silicon including microprocessors, other digital circuitry and various types of analog devices. Research at Caltech, the University of Michigan (U. Mich), and other academic/industrial fabrication facilities around the world are constantly adding to the body of knowledge of how to construct novel devices out of one of the earth's most abundant minerals.

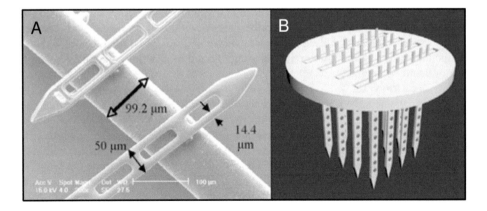

Figure 2.8: Examples of silicon-based technology developed at the University of Michigan's Center for Wireless Integrated MicroSystems. (A) lattice recording arrays compared to a human hair [77, 78]. (B) concept drawing of the lattice arrays used in a zero-rise recording array [77, 79]. *Photos courtesy of K. Wise.*

The brain and spinal cord is surrounded by cerebro-spinal fluid (CSF) contained in the arachnoid space located between the pia-mater[8] and the arachnoid[9]. The brain is not anchored to the skull at every point about its perimeter instead it is partially afloat in a sea of CSF. Thus, as we walk, run, and go about our everyday activities the CSF acts as a kind of shock absorber. This means that the brain is not absolutely stationary with respect to the skull. Micro-motion of the brain can cause problems with chronically implanted, rigid micro-electrodes that are anchored to the skull. While silicon is somewhat flexible when made into very thin structures, a more flexible substrate would be highly desireable. For example, a rigid microelectrode implanted in the brain might be integrated with a flexible micro-cable which connects it to the electronics attached to the skull. Toward this end, researchers are developing microelectrodes and cables constructed from flexible, biocompatible polymers such as parylene [80, 81, 82] and polyimide [83, 84].

At the University of Michigan's Center for Wireless Microsystems, researchers are developing a number of novel devices for use in neural interfacing applications. The center is headed by Dr. Kensall Wise who has over 30 years of experience with implantable microsystems. Among the center's most recent achievements are high-density cochlear electrodes for hearing prostheses, microelectrodes with integrated MEMS fluid handling systems, thin-film diamond probes for electrochemistry applications, and wireless implantable microsystems for multi-channel neural recording.

2.1.2.4 In Vivo Electrochemistry (IVEC)

There are basically two methods for detecting neurotransmitter release in living tissue: Microdialysis which relies on High Pressure Liquid Chromatography (HPLC) and In Vivo Electro-Chemistry (IVEC). Micro-dialysis provides excellent quantitative accuracy but poor temporal resolution due to the need to move fluid through tubing from the implant site to an HPLC apparatus for the analysis. Although it is an important clinical and experimental tool, microdialysis is too bulky (at present) to be useful as a self-contained, implantable neural interface. IVEC systems coupled with inexpensive, mass-manufactured microelectrode arrays have the potential to meet the criteria needed to produce small, implantable chemical detection systems.

The technique of in vivo electrochemical analysis was developed by Dr. Ralph Adams in the early 1970's [85] and has become a vitally important tool for the real-time detection and quantitation of neurochemicals in living tissue. Voltammetry refers to analytical methods in which a small amount of the analyte (the material being analyzed) is oxidized and/or reduced by an electrolytic reaction. The term "Oxidation" refers to the loss of electrons by a molecule or atom while the term "reduction" refers to a gain. Reduction/oxidation reactions are collectively referred to as redox reactions. During a redox reaction the movement of electrons sets up a measurable current that is proportional to the amount of the analyte that was electrolyzed.

In a voltammetry system there are two electrodes: the working electrode and the reference electrode. The working (a.k.a. recording) electrode is where the detection occurs. A voltage potential is applied to the reference electrode whose character and amplitude depends on the method of analysis (Fig. 2.9-A, CA, CV) and material being analyzed, respectively. An analyte that is electroactive will undergo a redox reaction in response to the applied voltage. The job of the reference electrode (and any auxiliary electrodes) is to allow the system to compensate for any drift that might occur in the potential applied to the reference electrode. The piece of equipment that monitors and compensates for any drift is the potentiostat. A simple diagram of a potentiostat and its connections are shown in Figure 2.9. There are a number of methods used in electrochemical detection and quantitation. The three that are the

---

[8] A thin highly vascular layer of connective tissue that is intimate with the folds of the brain.
[9] A delicate layer of connective tissue located under the dura-mater.

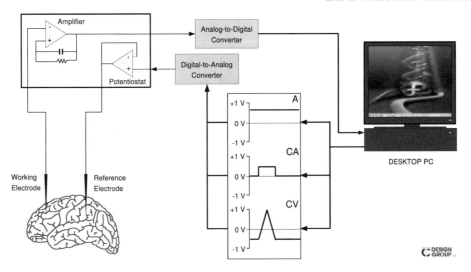

Figure 2.9: Block diagram of an IVEC system. A desktop PC is used to generate the waveform which is applied to the working electrode. The output of the D/A converter is sent to the potentiostat (shown here as a unity gain buffer). The current detected by the working electrode is amplified and converted into a proportional voltage (trans-resistance amplifier). The output waveform is digitized and processed in the PC. A: amperometry, CA: Chronoamperometry, CV: Cyclic Voltammetry.

most widely used for *in vivo* studies are: cyclic voltammetry, chronoamperometry, and amperometry. Differences between the three techniques include the shape of the waveform applied to the reference electrode (Fig. 2.9-A, CA, CV) and the way the output recorded at the working electrode is processed to compute the analyte concentration. Each technique has its own unique set of advantages and applications which are to lengthy to discuss here. Instead I will refer you to "Electrochemical Methods" by Bard and Faulkner [86] for a more detailed discussion on the subject.

The electrodes used for IVEC range from single carbon fibers to microelectrode arrays fabricated on a number of different kinds of substrates[10]. The ideal situation from a neural interfacing perspective would be to have a "lab-on-a-probe" which would be capable of detecting the simultaneous release of different neurotransmitters and the corresponding electrical activity. Realization of this goal depends on the fabrication of probes containing multiple microelectrodes and the ability to apply different coatings to each of these sites. Typically, the microelectrodes themselves tend to be fabricated from platinum although other materials have also been used. In the past, ceramic, and silicon were the most prevalent substrates for electrochemistry probes. While still widely used, more recent work has focused on flexible polymer substrates such as parylene, exotic silicon variants, diamond, saphire, etc. Generally speaking, the higher the dielectric constant of the substrate the better it will be for electrochemistry, although dielectric properties are not the only consideration. For example, microelectrode arrays fabricated on silicon, a semiconductor, have also been used for IVEC with good results (Fig. 2.10). Silicon substrates

---

[10]The base material from which the probe is fabricated.

offer the advantage of being able to fabricate micro-electronic devices directly on the probe using robust and widely available manufacturing techniques.

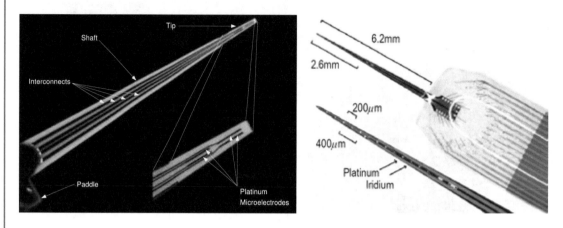

Figure 2.10: Examples of microelectrode arrays used to record *in vivo* electrochemical activity. (Left) 4 site, ceramic substrate array developed at the University of Kentucky's Center for Microelectrode Technology. (Right) 16 site, silicon substrate recording array developed at the University of Michigan's Center for Neural Communication Technology. *Left photo courtesy of P. Huettl; right photo courtesy of D.R. Kipke.*

Not all analytes are electroactive. That is, some molecules hold on to their electrons more tightly than others. In this case, microelectrodes must be "functionalized" with enzymes or other substances which convert the molecule of interest to a form that can be detected. One such example is the neurotransmitter Glutamate which requires the use of the enzyme Glutamate Oxidase to convert it into an electroactive form. As such, IVEC is a less direct method of detection than HPLC. In the case of an enzyme-assisted detection strategy, the output of the working electrode is a current that is proportional to the amount of the analyte that was converted to an electroactive form and subsequently oxidized/reduced. The redox current data is then used to compute the concentration of analyte present at the surface of the working electrode at the time the electrolytic reaction took place. Studies have demonstrated that the quantitation achieved with enzyme assisted IVEC is comparable to similar methods utilizing HPLC. As such, IVEC is becoming the method of choice for rapid, *in vivo* analysis of neurotransmitter release and is particularly well suited for neural interfacing applications.

At the time of this writing there are several companies offering IVEC systems for use in research. Quanteon, LLC was started by Greg Gerhardt (a former Ph.D. student of Ralph Adams) in 1995 to develop and market systems for *in vivo* electrochemisty. The system (Fig. 2.11) sold by Quanteon performs IVEC using amperometry and chronoamperometry. Mark Wightman (a former postdoc of Ralph Adams) of the University of North Carolina has developed IVEC systems which utilize cyclic voltammetry primarily for the detection of dopamine. IVEC equipment like that developed by Dr. Wightman is available from Cypress Systems[11]. Both Drs. Gerhardt and Wightman have dedicated

---

[11]A division of ESA Biosciences.

a large part of their careers to the development of the techniques and technology which make IVEC possible.

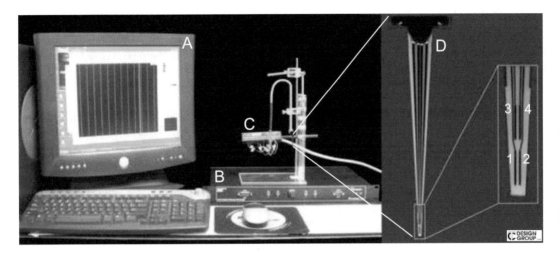

Figure 2.11: F.A.S.T. IVEC System by Quanteon LLC of Nicholasville, Kentucky shown with a typical microelectrode array attached. The system utilizes PC based data acquisition and measurement software (A). The system shown is capable of recording from up to 8 channels simultaneously *in vivo*, with a temporal resolution of less than 100 milliseconds. It is able to measure a variety of neurochemicals in living tissue such as those listed in Table 1.1 using either amperometry or chronoamperometry.

A relative newcomer to the IVEC scene is the University of Michigan's Center for Wireless Integrated Microsystems and the Center for Neural Communication Technology. There, a research effort lead by Drs. Daryl Kipke and Richard Brown is developing silicon-based, multi-channel probes (Fig. 2.10-B) and a system capable of simultaneously recording electrical and neurochemical (dopamine, choline, or serotonin) activity. Building upon the center's expertise with semiconductors, their current probes are fabricated on silicon substrates and will incorporate on-chip electronics. The group has plans to begin fabrication of high channel count devices using flexable polymers in the near future. If successful, theirs will be the first system capable of performing *in vivo* electrochemistry and electrophysiology using the same probe and a single software package.

2.1.2.5 Chemical Stimulation

The ability to stimulate a synapse or a small group of synapses using neurochemicals relies on the ability to deliver the chemical in question with extraordinarily high precision. At present chemical stimulation is accomplished using glass micropipettes. The pipettes are tapered to an extremely fine point using a pipette puller, filled with the desired solution, then piggy-backed on a microelectrode. The untapered end is fitted with flexible tubing connected to a Picospritzer[12] which provides very precise pressure to eject small quantities of the liquid from the tapered end of the pipette. When the micro-electrode/micro-pipette combination is placed in neural tissue it provides a means by which the electrical (or chemical)

---

[12]Picospritzer is a registered trademark of Parker Hannifin Corp, Pine Brook, NJ.

response to the ejected substance can be recorded. Compared to the size of a synaptic terminal which averages less than 1 micron, the probe/pipette combination is several orders of magnitude larger. Although this method is considered state-of-the-art at present, it is of course at best. Researchers at a number of universities around the world are developing Micro Electro Mechanical Systems (MEMS)[13] capable of handling nanoliter quantities of liquids. These miniature fluid handling systems, called micro-fluidics, hold the promise of one day allowing us to construct probes which can not only record chemical activity but release a variety of neurotransmitters on command. If micro-fluidics could be coupled with an onboard repository of living neuro-secretory cells the device could have an inexhaustible neurochemical reservoir. Otherwise, there will have to be an external means of refilling the device when its chemical supplies are exhausted.

There is presently no technology that will allow us to target a chemical stimulus with the precision of a single synapse. This is because we do not yet know how to fabricate (or place) devices on a scale which would permit such fine control. However, this fact does not negate the usefulness of less precise devices. Neuromodulators, which play a crucial role in neural processing, are released into the extracellular space and diffuse to the surrounding synapses. These neurochemicals effect the spiking activity of neurons proximal to the site of release. Dysfunction of this neuromodulatory system has been implicated in a number of conditions including certain forms of mental illness. Ergo, a device capable of even diffuse release of neurochemicals could prove highly useful for diagnosing and treating such disorders.

2.1.3 BIO-HYBRID CIRCUITS, NEUROCHIPS, AND NEUROPROBES

The ever-shrinking dimensions of what we can build will eventually allow the fabrication of arrays containing thousands of microelectrodes. These ultra-dense arrays will contain integrated processing circuitry to permit wireless communication with computers. Such arrays will permit the study of living neural networks with unprecedented detail. Cell biology is a field of study concerned with understanding cell growth, development, physiology, and proliferation. Cell culture refers to both a discipline and a body of knowledge regarding how cells can be maintained *ex vivo*, induced to proliferate and differentiate into specialized tissues in the laboratory. Through the combination of silicon micro-electrode arrays, cell biology, and cell culture, it is now possible to grow intact neural networks *in vitro* directly on silicon microchips. Development of these high-tech tools began in the Laboratories of Jerome Pine at the California Institute of Technology in the U.S. and Peter Fromherz at the Max Planck Institute of Biochemistry in Martinsried, Germany.

Dr. Fromherz, building upon his earlier work [87], developed a technique whereby the outgrowth of leech neurons could be guided along pre-determined pathways to form defined arborizations [88] (Fig. 2.12-A). These early experiments utilized glass coverslips as the substrate and extracellular matrix protein as the patterning agent. Neurons grown using this technique were then used to evaluate the hypothesis that dendritic arborizations were capable of performing nontrivial processing on their inputs. At the time the idea of dendritic processing was the subject of considerable debate.

To test the hypothesis that dendrites could be more than just passive conductors the XOR[14] function was chosen. Many artificial neural network models of the time relied on what is known as the McCulloch-Pitts neuron. This type of artificial neuron was modeled on the theory that dendrites of real neurons are passive structures which perform a simple summation of their inputs. McCulloch-Pitts neurons are not capable of implementing the XOR since to do so requires the ability to perform an

---

[13]MEMS research can be followed in the IEEE/ASME Journal of Microelectromechanical Systems (http://www.asme.org) and at the annual IEEE International Conference on MEMS (http://www.ieee.org).

[14]The output of the XOR function is false when the inputs are either all true or all false: otherwise the output is true.

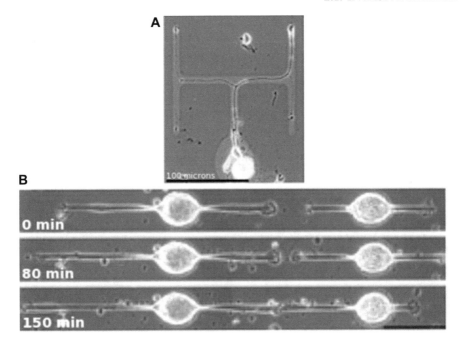

Figure 2.12: Results demonstrating patterned outgrowth of a leech neuron (A) [88] and a sequence of three photos demonstrating guided outgrowth and synapse formation between two neurons (B) [89]. Scale bars are 100 μm. *Photos courtesy of P. Fromherz.*

inversion[15] operation. Thus, the XOR was attractive as a test since it is one of the operations that cannot be performed by the theoretical "point neuron" of McCulloch-Pitts style artificial neural networks and should not be able to be performed by a real neuron if the current theory was correct. In 1993, the results of these experiments were published and demonstrated not one but four different ways that the XOR function could be implemented in the patterned arborizations [90] of living neurons.

In 1995, Fromherz and Stett demonstrated that leech neurons grown on specially designed silicon substrates could be capacitively stimulated using microscopic, metal-free, field effect transistors (FETs) [91]. By making the FETs metal-free the concern of leaking toxic metals into the extracellular medium and slowly poisoning the cells could be avoided. The neuron's cell membrane acts as one plate of a capacitor separated by the silicon dioxide insulating layer from the p-doped layer of the FET which acts as the other plate (Fig. 2.13). Stimulation occurs when the electric field coupled across the silicon-neuron junction achieves sufficient charge density to depolarize the voltage gated ion channels in the neuron and produce an action potential. In such an arrangement no faradic current flows which eliminates possible toxic effects of repeated stimulation. Stimulation using current can cause leaching (migration) of cytotoxic materials from the substrate into the surrounding extracellular space effecting the health and longevity of the neurons. Also, high currents can damage the cell directly. The following year both recording and stimulation were demonstrated using capacitive coupling of the silicon-neuron junction [92].

---

[15]An inverter is a basic logic element where the output is the opposite of the input.

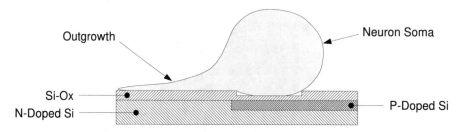

Figure 2.13: Illustration of the metal-free silicon-neuron junction pioneered by Peter Fromherz. *Based on [91, p. 1670, Fig. 1].*

The intervening years between 1995 and 1999 saw improvements in techniques, development of better methods for visualization of living neural networks, further characterization of the silicon-neuron junction, and development of new structures to retain the neurons on the silicon substrates. In 1999, this work produced the first demonstration of the formation of a functioning electrical synapse between two neurons using guided outgrowth (Fig. 2.12-B). Later in 2001, Dr. Fromherz's and Günther Zeck[16] published an article [93] describing a new methodology using a picket fence type of structure constructed from parylene to retain the neuron cell body in place over the active electronics. In previous experiments the cell bodies tended to migrate away from their initial positions over the FETs which caused problems with stimulation and recording. The new structure prevented migration of the soma while permitting the neural processes to grow un-hindered (Fig. 2.14). A later article [94] described the Electrolyte-Oxide-Silicon transistor which permits the electrical activity of the neuron to directly control the source-to-drain current of the FET.

Dr. Fromherz's has continued to advance the study of silicon-to-neuron devices and has added to his repertoire the construction of multi-transistor arrays which provides, "... time-resolved images of evoked field potentials and allows the detection of functional correlations over large distances"[95]. The arrays are capable of imaging electrical activity at exceptionally high resolution (16,384 pixels). Fromherz uses these devices to study patterns of activation in cultured hippocampal[17] slice preparations.

In the late 1970's, Dr. Jerome Pine of the California Institute of Technology (Caltech)[18] was developing a technique to record from cultured neurons using planar, extracellular microelectrode arrays. The arrays consisted of 32, 8μm x 10 μm thin film microelectrodes patterned on glass and positioned at the bottom of specially prepared petri dishes. Neurons obtained from rat superior cervical ganglia were cultured atop the arrays for 8 days after which electrical activity could be recorded using the array. Pine speculated in his 1980 paper on this work [96] that such cultures could, "...create neuronal networks which offer the potential for studies which are now technically difficult or impossible to perform *in vivo*." The technique also offered the ability to record from multiple neurons simultaneously without damaging the cells, thus enabling long-term studies of development and plasticity.

---

[16]Max Planck Institute for Biological Cybernetics in Tübingen.
[17]The hippocampus is a structure in the forebrain that is involved in memory and spatial navigation.
[18]While on leave at the Washington University School of Medicine.

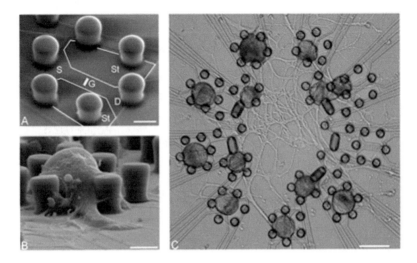

Figure 2.14: Biohybrid silicon-neuron technology from Dr. Fromherz's Laboratory. (A) EM of picket fence made of polyimide. Stimulator wings (St) and electrolyte-oxide-silicon (EOS) field-effect transistor (S source, D drain, G gate) are shown. Scale bar = 20 μm. (B) EM of snail neuron in a picket fence after three days in culture. Scale bar = 20 μm. (C) Micrograph (viewed from above) after 2 days in culture, showing neurons held in picket fences on a circle of two-way contacts. Neurites (bright threads) can be seen connecting the neurons forming a functioning neural network. Scale bar = 100 μm. *Photos courtesy of P. Fromherz* [93].

After returning to Caltech, Pine continued to develop this recording technique and began work with Dr. David Rutledge[19] and Wade Regehr[20] to develop a new silicon-based technology for use with cultured neural networks. In 1988 The trio published their work [97] describing a micromachined "diving board" electrode which extended from a pedestal and made contact with the top of the cell in the petri dish. While the work was successfully concluded with the demonstration of stimulation of- and recording from the cultured neurons, the technology needed improvement especially with respect to the seal obtained between the electrode and the cell.

In the early 1990's Dr. Pine formed a collaboration with Dr. Yu-Chong Tai of Caltech's Micro-machining Laboratory to continue the work on applying microelectrode arrays and silicon substrates to the *in vitro* study of living neural networks. The technology developed from this collaboration would later come to be known as the "Neuro-probe" and the "Neuro-chip." The basis of both devices were microfabricated "neurowells" into which immature neurons were placed. Grillwork at the top of the well prevented the soma from migrating out but allowed the axons and dentrites to grow out of the well onto the top of the chip where they could synapse with processes from other "caged" neurons. A gold electrode at the bottom of each well-permitted recording and stimulation.

---

[19]California Institute of Technology.
[20]Ph.D. candidate, Caltech.

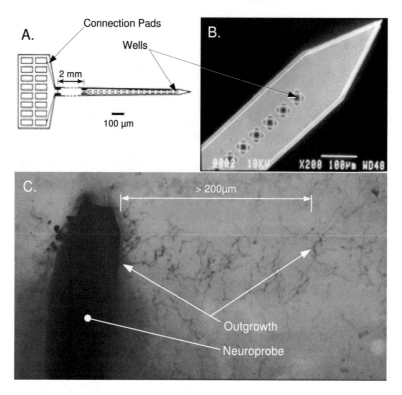

Figure 2.15: The Neuroprobe. (A) diagram showing the layout of the neuroprobe. Circuitry (not shown) connects each of the 16 pads on the paddle with their corresponding microelectrode at the bottom of each of the 16 wells on the shank. (B) electron micrograph of the tip of a completed neuroprobe prior to loading with neurons. (C) *In vivo* results: Light micrograph of stained [98] neurons showing neurite outgrowth from the wells into rat brain tissue. *Diagram and photos courtesy of J. Pine* [99].

The neuroprobe (Fig. 2.15) was a dagger-like, linear array of neurowells with a paddle shaped connection area at the top. The shape of the neuroprobe allowed it to be loaded with neurons then the shaft inserted into brain tissue. Over the course of the next few days the neurons grew their processes out of the well and into the surrounding neural tissue. Experiments were performed first with cultured brain slices and then *in vivo* in rat hippocampus. Both studies demonstrated that the cells in the neuroprobe were capable of surviving, growing, and sending out processes that are incorporated into surrounding brain tissue. All the probes used in the studies lacked electrodes and as such were not capable of recording or stimulation. The primary goal of the study was to test the effectiveness of the well design for containing the cells, demonstrate survival of the neurons and outgrowth of neural processes into the surrounding tissue. All of the projects initial goals were met and the next step was to fabricate probes with electrodes for testing. Unfortunately, this work was not continued much past the mid 1990's due to factors unrelated to the efficacy of the technology. However, work on the Neurochip has progressed significantly since this time.

The first version of the Neurochip utilized the same well design used on Neuroprobe. The neurochip is an *in vitro* device intended to serve as a substrate on which networks of cells could be grown and studied much like Pine's earlier work with thin-film microelectrodes. Early versions of the neurochip featured wells etched into bulk silicon [100] and enabled Pine's group to demonstrate viability of the cells and show proof-of-concept of the design. The neurochip allowed them to engage in the long-term study of both the development and activity of live neural networks. Despite these successes, the early neurochip suffered from several problems, primarily: the wells were difficult to fabricate in large numbers and the neurons sometimes lost connection with the electrodes as they grew out of the top of the well.

The second generation of the neurochip was implemented using cages constructed of parylene (Fig. 2.16-A), a flexible polymer. The new design incorporates advances in materials, fabrication techniques, and lessons learned from the first generation chips. These improvements include changes in the cage design to better contain the cell body and to maintain electrical contact. Rather than have the soma sit at the bottom of the well and its processes grow out the top, the second generation uses tunnels extending from the side of the cage. As the neuron sends out processes through these tunnels they help retain the soma in place over the microelectrodes. This ingenious design along with the method of fabricating parylene structures on silicon was arrived at though years of trials by Pine and Tai. The cages themselves are anchored into the silicon using vertical holes that extend deep into the substrate. The process by which the holes are constructed was pioneered in Dr. Tai's lab.

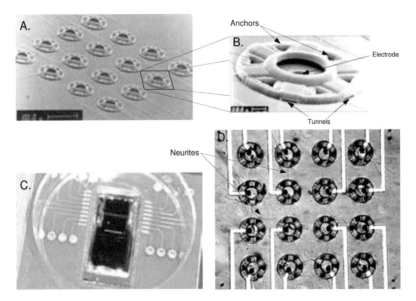

Figure 2.16: The Neurochip 2007. (A) SEM of a 4 x 4 array of parylene wells fabricated on silicon substrate. (B) Close-up of one of the wells. (C) the assembled, mounted and connected neurochip shown positioned at the bottom of a petri dish. (D) Photo of neurites growing from the wells and forming synapses on the surface of the silicon [101]. *Photos courtesy of J. Pine.*

One of the problems encountered with the use of the neurochip was the time it took to load the neurons into the cages. The second generation neurochip contained a 4 x 4 array of neurocages but the plan was to fabricate much larger arrays to permit far more complex networks to be grown and studied. A new method had to be found which would permit rapid and perhaps semi-automatic loading. This required a new approach to cell handling and placement. The solution to this problem came in the form of optical tweezers built by Gary Chow [101] to pick up and place the neurons into the neurocages.

Whenever light strikes an object and is refracted (changes direction) it imparts a small momentum to the object; optical tweezers make use of this phenomena to pick up and move cells. When a laser beam is focused onto a transparent refractile sphere (the neuron), the beam is bent as it enters the cell due to differences in the refractive indices of the cell and surrounding medium, and bent again as it exits on the other side. Due to the opposing forces exerted on the cell by each bend in the beam, the cell is held at the beam's point of focus. By moving the laser beam (and the cell along with it) the "captured" cell can be maneuvered into a neurocage. Optical tweezers have been used in a number of other applications [102] including manipulation of other type of cells [103], airborne particles [104], and protein translation [105]. To minimize the chance of damage to the cells, an infrared[21] laser was used. Future Neurochip designs will consist of 60 (or more) cages and permit the study of large *in vitro* neural networks. The optical tweezer technology enables the rapid loading of these high density neurochips.

The work being conducted by Drs. Pine and Fromherz is creating new avenues of research that will help us to better understand the functioning of biological neural networks. The ability to grow neurons on active substrates capable of recording and stimulation will permit the study of biological neural networks in ways that would never be possible in a living organism. It is difficult to study the details of such networks *in vivo* for a number of reasons including:

- The physical placement of electrodes into living brain tissue may disrupt fine connections we wish to study.

- Networks in the brain are highly interconnected making isolation and study of smaller interconnected groups of cells impossible.

- We presently lack tools able to visualize synaptic morphology *in vivo*[22]. Current research indicates that the size of the synapse may have a role in information processing and changes in size may have a role in age-related mental degradation [106].

Tools like those being developed by Drs. Pine and Fromherz combined with other technologies could overcome this problem by permitting the detailed electro-chemical study of both the function of the network as a whole and the individual neurons of which it is comprised. Furthermore, networks grown in petri dishes can be periodically placed under a microscope and their physical properties observed; something that is not possible with a living brain.

With regard to brain-machine interfaces, if the theory that each brain is truly unique with respect to how we process and store perceptual experiences, an adaptive interface will be required. Some future device like the Neuroprobe that uses biological neurons to make the connection between the brain and the electronics, might enable the requisite adaptive, direct neural interface. Using some training paradigm the cells could, as they grow and connect, learn to interpret the activity of the surrounding neural network into a form that could be used by the connected computer system. There are a number of advantages to this strategy (if it could be made to work) over using microelectrodes:

- Neurons are capable of making far more precise and well-defined connections than are possible now or for the foreseeable future with MEAs, microwires, or other inorganic means.

[21]Living cells are semi-transparent to infrared radiation.
[22]The average size of a synapse is approximately 0.5 μM compare that to a human hair which is about 100μM.

- Neurons have the capacity to adapt and learn, thus providing the possibility of signal processing and interpretation built into the substrate of the interface.

- Neurons are not susceptible to electrical noise like microelectrodes.

- The connections made by the neurons might be better tolerated over the lifetime of the person than nonbiological/inorganic materials.

- Future advances in other fields such as genetics and cellular/molecular biology may allow the engineering of a type of "interface neuron" specifically designed to connect human-made circuitry to *in vivo* neural networks.

Neural interfaces of the future will probably look far more "organic" than those envisioned today. Such interfaces may contain a core of electro-optical processing circuitry surrounded by human-engineered cells. The hybrid device would then be implanted into a damaged brain area where it would grow and integrate with the remaining tissue. The organic and inorganic components would work together, interacting with the existing brain tissue, to "learn" how to replace the lost functionality. Such a device is far in the future indeed but the technologies needed to realize it are already in the early stages of development. The creation of such a device relies on expertise from many disciplines. The work of Drs. Fromherz and Pine are but two examples of the highly interdisciplinary nature of neural interfacing/enginering.

2.1.4   FUNCTIONAL ELECTRICAL/NEUROMUSCULAR STIMULATION

Functional Electrical Stimulation or FES is useful when the muscles are intact but signals cannot reach them due to damage to the nerves and/or the spinal cord [107]. FES systems control muscle movements using stimulating electrodes located either on the skin above the muscle [108] or implanted directly in the muscle tissue [109]. Present systems such as the Parastep®[23] use a remote keypad to select pre-defined muscle activation patterns to, for example, stand from a sitting position and/or walk using a walker.

Dr. Gerald Loeb of the A.E. Mann Institute for Biomedical Engineering and pioneer of the BION®system has been studying electrode construction, spinal reflexes, EMG[24], and FES for decades. His laboratory is developing wireless, implantable micro-stimulators which can be injected into muscles using a large diameter needle. These tiny implants are inductively powered and controlled from a small computer worn by the user. Another research group headed by Drs. Ranu Jung and Jimmy Abbas at the University of Arizona's Center for Adaptive Neural Systems, is developing systems and methodologies using FNS (Functional Neuro-muscular Stimulation) and other paradigms to re-animate paralyzed limbs. Their current efforts include the development of a MEMS-based neural clamp that can be reversibly attached to spinal roots. The clamp would be used as the mechanism to hold micro-electrodes in place for the purpose of stimulating peripheral efferents to produce muscle contraction.

FES/FNS systems not only help to restore limited movement to paraplegics, but help prevent muscle atrophy which occurs due to disuse. Some day we will be able to build the "ultimate" system which bypasses the injured spine and permits direct mental control of the muscles. This might be done using the combination of a brain-to-machine interface and a system to translate detected brain activity into movement commands. The commands would then be converted into the correct sequence of stimuli to contract the muscles and carry our the persons intended movement.

---

[23]Parastep is a registered trademark of Sigmedics Inc., http://www.sigmedics.com.
[24]ElectroMyoGraphy: recording the electrical activity in muscle using surface electrodes.

As I read about these systems I cannot help but be reminded of the old Star Trek episode "Spock's Brain." In this episode the brain of the ill-fated Vulcan is stolen and his friends must bring his body along as they search for the missing organ. Instead of carrying his body on a stretcher the Doctor constructs a device (controlled by a wireless, handheld remote) to move Spock's muscles and make him walk. I remember joking as I watched that episode years ago, "Gee, it's the 23rd century and the best they can do is have him move like Mr. Roboto? They should have him break dancing by then." A "few" years and a Neuroscience Ph.D. later, I realize that having a brain-less body do anything at all would be pretty amazing.

2.2   NON-INVASIVE NEURAL INTERFACES

This category of interfacing strategies does not require surgery, injection or any other type of procedure which "invades" the internal spaces of the body. These technologies (if perfected) would be useable by any person without bodily alteration. This includes devices or electrodes that can be worn, adhered to the skin, or otherwise nonpermanently attached to ones person. Such devices have the potential advantage of not requiring specific licensing or approval for use by normal individuals other than ensuring electrical and mechanical safety. Thus, noninvasive neural interfaces, in all likelyhood, will be the first type of interface that shows up on the store shelves for you and I to use. These technologies are not currently fast enough to replace a computer keyboard and mouse. So, unless you are a quadriplegic and have no other option it is not likely that you would want to use them in their present state of development; In the future this will surely change.

2.2.1   ELECTROENCEPHALOGRAPHY (EEG) AND
MAGNETOENCEPHALOGRAPHY (MEG)

Experimental, EEG-based neural interfaces often utilize scalp electrodes affixed to a type of snug-fitting, fabric cap that can be worn by the user. The location and spacing of the electrodes is usually according to one of the standard systems of placement such as the "10-20" system [110]. The electrodes are used to detect fluctuations in the electric field produced by the brain. The field is a composite of all the electrical activity emanating from all the neurons at a given instant in time. Deconvolution of this activity into something useful is similar to the "cocktail party" problem discussed earlier where each electrode sees a different "view" of the composite activity.

Use of EEG as the basis of a neural interface depends on some very sophisticated and computationally intensive calculations to extract meaningful control information from the composite activity. EEG BMIs have improved remarkably over the years both in accuracy and speed. This increasing rate of success is directly tied to the increase in processing "horsepower" available on desktop PC's and their decreasing cost. Deconvolution of EEG is a computationally bound problem, that is, a problem which relies on the ability to perform many calculations at a high rate of speed. As computer technology progresses and the systems become smaller, faster, and less expensive, EEG-based BMI will likely become practical as a way for everyone to interact with machines.

Drs. Jonathan Wolpaw and Dennis McFarland are two well-known names in the field of electroencephalography (EEG). Their work on building an EEG based brain-computer interface [5] has spanned nearly 20 years, produced scores of journal articles, and led to the open-source BCI-2000 system [111][25]. In Europe and Asia there are several efforts to develop EEG-based brain-computer interface systems. The Berlin Brain-Computer Interface project at the Fraunhofer Institute directed by Drs. Klaus-Robert

---

[25]http://bci2000.org

Müller, Gabriel Curio and Benjamin Blankertz [112, 113] has been on-going since 2000. They have made a number of advances in EEG-BCI especially in producing systems which require minimal user training. Also in Germany at the Institute of Medical Psychology and Behavioral Neurobiology at the University of Tübingen, researchers are working to decode hand movement from EEG signals to allow control of a robotic limb [114]. In Switzerland at the IDIAP Research Institute, Dr. José Millán and his team have developed an EEG-based system which permits a user to mentally guide a robot around an indoor maze [115]. At the Brain Science Institute in Japan the work of Dr. Andrzej Cihcocki is helping to improve the speed and accuracy of EEG-BCIs by developing novel algorithms to extract pertinent features from human EEG signals [116].

Magnetoencephalography (MEG) provides similar information to EEG except the detected magnetic field is 90° rotated from the electric field. That is, the electric field detected by the EEG is perpendicular to the skull while the magnetic field is parallel to the skull. The equipment needed for MEG is very bulky owing to the super-cold temperatures needed by the array of ultra-sensitive, superconducting quantum interference devices (SQuIDs). SQuIDs are capable of detecting the tiny magnetic fields produced by an action potential. Since EEG can deliver essentially the same information but with far less bulk and expense it is currently preferred over MEG as the method of choice for noninvasive interface systems.

2.2.2   ELECTROMYOGRAPHY

Electromyography or EMG involves the placement of electrodes on the skin over a muscle whose activity one wishes to monitor. The contractile state of the muscle is reflected by the detected electrical activity and can be used to control a prosthesis. EMG has been used to control various type of myo-electric prosthetics with good results. However, it cannot deliver the fine motor control and sensory input of the natural limb. The first prototype of the prosthetic arm for the DARPA RP-2009 project (discussed in Sec. 3.3) used this type of control scheme. There have been a number of research efforts to use EMG to interact with virtual environments including a hand model [117], neck muscle EMG to track head position [118], a system for gesture recognition [119] and the just-mentioned DARPA project.

2.3   TECHNICAL CHALLENGES TO IMPLANTABLE DEVICES

One of the significant challenges to neural interface development is designing technologies which can withstand years of implantation in the body. The internal environment of the body is highly ionic (salty) and will wreak havoc on most materials commonly used in the production of semiconductors and electrodes. Not only does the device have to survive this corrosive environment but it has to remain chemically inert to prevent damage to the surrounding tissue. There is a need to find coatings for present materials or alternative materials which will allow implants to withstand the harsh environment of the body for decades. Much of the neural interfacing work conducted thus far has utilized microfabricated components constructed from silicon. Since it is highly desirable from both a cost and R&D standpoint to utilize microchip manufacturing techniques, methods for protecting the fragile electronics must be found. Additional issues for chronically implanted devices include:

- Heat: implanted, active devices generate a small amount of heat during powered operation. As devices become more complex, smaller and circuitry density increases, more heat will be generated. The surface temperature of the device should not exceed body temperature to prevent possible undesirable effects to the surrounding tissues.

- Size: the devices must be small enough so as not to interfere with movement or other activities.

- Aesthetics: scarring, devices visible as lumps under the skin, etc.

2.4   POWER AND COMMUNICATION

The final topic to be covered in this section is how implanted devices are powered and how they communicate with the "outside" world. In the past, these two items were accomplished using a hardwired connection. That is, a multi-conductor electrical socket mounted to the skull or elsewhere that was connected to the external equipment. Though this type of arrangement is still used due to its low cost and simplicity, wireless technologies are gaining in popularity for a number of reasons:

- Less chance of infection since everything is contained under the skin.

- Decreasing cost: always a big factor with limited research budgets.

- Increased availability: a number of companies now offer out-of-the-box solutions so researchers no longer have to spend the time to develop their own systems.

- Decreased size.

- Decreased power requirements: in the past, large battery packs have limited the usefulness of un-tethered systems.

- In awake, behaving subjects it is difficult to study natural behaviors if the subject is in the unnatural situation of being tethered to a rack of equipment by a cable.

Long-distance, wireless transmission of data is possible utilizing a number of standard methods including Amateur (HAM) Radio, WiFi, cellular, and/or satellite communication channels. In the U.S., the band of frequencies utilizable for long-range communication is strictly regulated and a license may required to operate on certain frequencies. Short-range communication devices are less strictly regulated and the frequencies assigned for these devices are (in general) available for un-licensed use. Many of the custom designed devices available for laboratory use, utilize short range transmitter/receivers which operate in the bands assigned for consumer products such as bluetooth and cordless telephones. These devices usually have a ranges from 2 to approximately 100 meters.

Implanted devices such as cardiac pacemakers are often powered using batteries that must be replaced every five years or so. Neural interfaces which operate continuously and transmit/receive data have much higher power requirements and cannot be powered solely from an implanted power cell. These types of devices are usually powered either one of two ways: (1) by a cord which extends through the skin to plug into a power supply, or (2) by an implanted rechargeable battery with inductive charging circuitry. In the case of the latter, the transmission of power occurs between two wire coils, one implanted under the skin the other lying on the surface of the skin. The two coils form an air-gap (or in this case a skin-gap) transformer where the magnetic field generated by the alternating current flowing through the exterior coil induces a current to flow in the internal coil. The current is then rectified into DC, regulated to the proper level and fed to the battery's charging circuit.

There are certain species of animals that are capable of generating natural electricity. The best known is *Electrophorus electricus* or the electric eel found in the Amazon. The electric eel can produce up to 800 watts of energy per discharge, enough to light a small Christmas tree. Perhaps some day a way will be found to produce an implantable biological power source based on this or other examples found

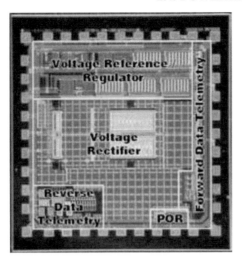

Figure 2.17: An example of the degree of miniaturization that can be presently be attained in wireless circuitry. The General telemetry module developed at University of Michigan's Center for Wireless Integrated MicroSystems provides both two way communication and power regulation. The module measures 2mm x 2mm [120]. *Photo courtesy of A. Sodagar.*

in nature. Such a device would resemble an artificial organ, connected to the blood supply and utilizing the nutrients found there to feed the cellular mechanisms that produce bioelectricity. There have been a few investigations into how bioelectricity might be harnessed for similar applications but (as far as I know) no practical methodologies for its use have yet emerged.

Present interface designs which have no trans-cutaneous (through the skin) connections use a combination of an impanted wireless transmitter/receiver and inductively rechargeable batteries. Systems such as these have proven very useful in research since they permit the subject complete, natural freedom of movement. Wireless systems have been used in both humans and animals since the 1960's and have been refined and miniaturized over the intervening years.

CHAPTER 3

Current Neuroprosthesis Research

"The technology for nonsensory communication between brains and computers through intact skin is already at our fingertips, and its consequences are difficult to predict. In the past the progress of civilization has tremendously magnified the power of our senses, muscles, and skills. Now we are adding a new dimension: the direct interface between brains and machines."

- Prof. José M.R. Delgado, MD/DSc [39, p. 95] -

Neuroprosthetics represent the amalgamation of many technologies and disciplines to create systems that, for example, permit the direct connection of a prosthetic arm to the brain. At the time of this writing, there are a large number of groups conducting neuroprosthetic research. This work is most often done in conjunction with work on neural interfacing since you need the interface for the prosthesis function. This chapter features highlights from some of those groups but should in no way be construed as the final word on the field as that changes from year to year. I have included contact information for the labs mentioned in this and the previous chapter at the end of the book. If you are interested by what you read here you should follow up by visiting the lab's web site or contacting them directly for more information. In the previous chapter, I discussed the various means by which the interface between the excitable tissue (muscle, nerve or brain) and a machine may be accomplished. This chapter focuses on current systems which use those interfaces.

3.1 VISION PROSTHESES

The entire process by which the visual world is converted to neural impulses, interpreted by the brain, and integrated into our perceptual experience is not well understood. However, there are parts of this process that are better understood than others and because of ongoing research more is being learned each year. The aim of artificial vision is to replace the damaged portion of the visual system with man-made components thus restoring some form of sight to the blind. There are three points in the human visual system where a neural interface might be established for the purpose of connecting a vision prosthesis: the retina, optic nerve, and the visual areas of the brain; prosthetics applicable to each area are discussed below.

The human eye functions much like a camera; rays of light pass through lenses (the cornea and lens) and strike the photo-sensitive film (the retina). Unlike the film in a camera, the retina is comprised of several interconnected layers of cells. The bottom-most layer contains the light-sensitive photoreceptors which convert an incident image into neural impulses. The layers above the photoreceptor layer (the light passes through the other layers to get to the photoreceptors) contain neurons which enhance/process the image. Thus, the light strikes the photoreceptors and is processed through the other cell layers until finally the neural data is conveyed to the brain via the optic nerve. The region of the retina which contains the highest density of photoreceptors (thus the highest resolution) is called the macula and is of significant importance in vision. The human eye contains two type of photoreceptors: cones which are involved in color vision and rods which are not sensitive to color but are more light sensitive than cones. The macula [leutea] contains a higher density of photoreceptors than the surrounding areas. It appears as a yellow spot on the retina owing to the large number of ganglion cells that can be found there. At the center of the macula is a slight depression called the fovea [centralis] which contains the highest density of cones and is responsible for the central part of our vision. As you move from the center of the visual field toward the periphery, the ratio of rods to cones increases while the overall density of photoreceptors decreases. Thus, we have the best sensitivity to light (night vision) with lower resolution in our periphery but the highest resolution and best color perception in the central part of our vision. This fact can be easily demonstrated in a near-dark situation by looking directly at an object then looking slightly off to the side while keeping the object in view. You will find that the object is more easily visualized by <u>not</u> looking at it directly. In

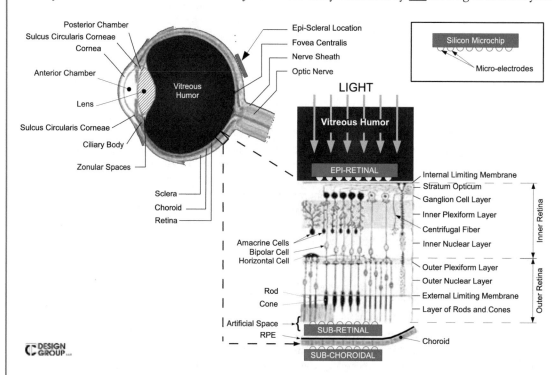

Figure 3.1: Basic anatomy of the eye and approximate locations for retinal implants. *Portions of this illustration were adapted from [7, Figs. 869 and 882].*

some individuals the photoreceptors in the macula begin to degenerate (termed macular degeneration or MD) leading to progressive loss of central vision. In others, the photoreceptors of the periphery (around the macula) begin to degenerate resulting in tunnel vision. This condition is termed Retinitis Pigmentosa (RP) and although it begins in the peripheral vision, it does eventually destroy the photorecptors of the macula as well. The end stage of both MD and RP is blindness. It is presumed that in individuals afflicted with these conditions that the other cell layers of the retina containing neurons remain intact and could continue to relay information to the brain if only there were a source of stimulation. In actuality, once the source of input (the photoreceptors) to the neural layers is lost, changes to those layers begin. Since a neuron's connectivity with other neurons is, in part, determined by its (and its neighbors) inputs, loss of those inputs results in undesirable changes in connectivity. Current research seems to indicate that artificial stimulation of those layers early in the course of the disease may help to maintain functional connectivity within the neural layer, protect the neurons from loss, and perhaps slow the degeneration of the photoreceptors [121, 122]. Since the artificial retina schemes discussed below rely on there being a functioning neural layer, preservation of the neurons is imperative.

3.1.1   SUB-RETINAL IMPLANTS

Sub-retinal implants are positioned in the eye in an artificially created space between the inner retina (containing the neurons) and outer retina (containing the photoreceptors). They have a slightly more complicated implantation procedure than epi-retinal implants but allow the stimulating portion of the retinal prosthesis to be placed in closer proximity to the inputs of the neural cell layer of the retina. Also, as the retina heals from the surgery, the implant is held in place by the surrounding layers of tissue. Thus, external means of restraint like tacks or a bio-adhesive are not needed (see epi-retinal implants below).

The first design for a totally self-contained sub-retinal implant was the Artificial Silicon Retina (ASR) conceived by Dr. Alan Chow in the mid 1980's and patented in August of 1989 [123]. The technology was further developed by Chow et al. and was the subject of numerous peer-reviewed articles [124, 125, 126, 127] and subsequent patents. The device consisted of a small array (2mm disc x 25 microns thick) of 5,000 micro-photodiodes which directly convert light into electrical pulses for stimulating existing retinal cells (Fig. 3.2). The early work on the feasibility of sub-retinal electrical stimulation and the refinement of the ASR design was carried out by Dr. Chow while at Loyola University of Chicago and his brother Vincent Chow, an Electrical Engineer [128]. In these experiments, strip electrodes were placed in the subretinal space of three adult Dutch Belted rabbits. Each electrode was connected to an external photodiode that, when illuminated, produced a current sufficient to stimulate the neural tissue of the retina. The results of these early experiments demonstrated that stimulation in the subretinal space could produce a response in the visual system. Furthermore, it provided support to the, "...idea that implantation of a photodiode based device into the subretinal space may provide a possible route for the restoration of visual function in patients blinded by photoreceptor degeneration" [128, p. 15]. Dr. Chow's work continued next in cats where the first prototype of the ASR microchip was tested. These experiments demonstrated that the microchip-based device would produce a response to light that could be measured in the neural tissue of the retina by recording electrodes placed on the eye. Some of the implants were allowed to remain in the cats and functioned for over three years, demonstrating the durability of the technology. Around the time of these experiments Dr. Chow and others formed Optobionics Corporation to further develop and commercialize the technology.

In January of 2000, the ASR entered clinical trials to assess the safety and feasibility of the implant. As part of this study, a total of 6 human patients suffering from retinitis pigmentosa were implanted. Since the photodiodes comprising the ASR were sensitive to both visible and infra-red light and the human eye is sensitive to only visible light, this provided a way to verify function of the ASR. That is, if a subject could perceive the sensation of light when the researcher illuminates the eye with an infra-red light source then the perception must be due to stimulation of the retinal cells by the ASR. Having the device implanted in human subjects who could directly report their experiences allowed the group to collect more definitive data on the functioning of the implant as a visual prosthesis.

In initial testing following implantation, four of the six patients reported the sensation of seeing light in response to infrared illumination of the ASR. This response slowly exhausted and became less noticeable by some patients in subsequent months. It was believed that certain characteristics of the

Figure 3.2: The Artificial silicon retina (ASR) pioneered by Dr. Alan Chow, MD and Vincent Chow, BSEE. The model shown measures 2 mm in diameter and 25 μm thick. It contains approximately 5,000 microphotodiode pixels each measuring 20 x 20 μm square. Left: ASR microchip shown atop a penny; Center: EM of the ASR microchip; Right: EM of the surface of the ASR microchip showing detail of pixels. The electrodes are the light colored, dimpled rectangles which are fabricated on top of the photodiodes (larger rectangles) [127]. *Photos courtesy of A. Chow.*

stimulation pulse caused undesirable changes to the tissue surrounding the ASR electrodes. This resulted in the stimulation current no longer being adequate to produce a response in the retinal cells. However, as the patients were monitored over the next 18 months they all showed remarkable improvement in vision, even in areas of the retina some distance from the implant. In an article published by the group in 2004, they indicated that, "Visual function improvements occurred in all patients and included unexpected vision improvements in retinal areas distant from the implant" [127, p. 469]. This led the group to the conclusion that, "...the observation of retinal visual improvement in areas far from the implant site suggests a possible generalized neurotrophic-type rescue effect on the damaged retina caused by the presence of the ASR" [127, p. 460]. In other words, the presence of the ASR in the sub-retinal space itself might be stimulating the damaged cells to produce neurotrophins (chemicals that stimulate growth in neural tissue) which lead to the rescue and repair of some of the retinal cells. Also, it has been shown by other researchers that certain types of electrical fields can affect cell growth, nerve repair, and retinal tissue preservation [129, 122]. This may mean that even though the microchip wasn't generating enough current to stimulate the retinal cells, it could be producing an electric field that had a positive influence on repair.

The data from the clinical trial complicated the picture of the efficacy of the ASR since it could not be clearly determined if the improvement was due to some neurotrophic rescue effect or a low-level electric field generated by the ASR or both. This result prompted the group to launch another study, this time in RCS Rats[1]. The results of the study suggested that in the rat, "...subretinal electrical stimulation provides temporary preservation of retinal function..." and "...implantation of an active or inactive device into the subretinal space causes morphological preservation of photoreceptors..."[121]. Later work [130] published in 2007 by a group working closely with Dr. Chow concluded that direct activation of the retina by the ASR induces activity in the superior colliculus[2] of RCS rats; indicating that it was indeed functioning as a visual prosthesis. Furthermore, the study demonstrated that there was, "... evidence for

---

[1]A breed of rat that develops a condition similar to RP in humans.
[2]An area of the brain responsible for orienting the eyes toward an intended visual target.

implant induced neurotrophic effects as a consequence of both it's presence and its activity in the retina." These findings put to rest the question of the ability of the ASR to function as a retinal prosthesis and indicate that it has additional efficacy as a means to slow or partially reverse photoreceptor degeneration.

The visionary work of Dr. Chow and his associates has inspired subsequent work by others on retinal prosthetics and demonstrated the safety, feasibility and durability of sub-retinal implants. While the story of the ASR continues to be written the work thus far bodes well for the treatment of photoreceptor degeneration in humans and it seems to only be a matter of time before a totally self-contained retinal prosthesis, powered only by incident light, will be realized.

Eberhart Zrenner, MD of the University Eye Hospital of Tübingen (Germany) and his colleagues are building on the work of Dr. Chow and others to design an externally powered, sub-retinal implant. The device is being developed in conjunction with his spin-off company Retinal Implant, AG[3]. The implant is a square microchip approximately 3mm x 3mm and about 50μm thick. It is fabricated on a flexible polyimide substrate and contains 1500 photocells connected to a 4 x 4 array of stimulating electrodes. The entire device is encapsulated in silicone elastomer. Unlike the ASR, the photocells do not produce the current used to stimulate the retinal tissue. Instead they gate the external power which is used for stimulation. This scheme insures that even in low light there will be sufficient power available for stimulation. Beginning in 2005, seven totally blind patients have undergone surgery for implantation of the Retina Implant device with promising results. Much was learned from these early human trials including several important technical modifications that needed to be made to the device. All patients were able to detect the location of light sources and sometimes identify simple shapes when presented with high foreground-to-background contrast (e.g., white on black) [131].

3.1.2 EPI-RETINAL IMPLANTS

Second Sight Medical Products, Inc.[4] was founded by Alfred Mann in 1998 to develop implants designed to help those blinded by retinal degeneration. Their first product, the Argus 16, entered clinical trials in February 2002 and to date a total of six subjects have been implanted with the technology. This experimental device consists of a 4 x 4 epi-retinal (Fig. 3.1)[5] array of electrodes that are attached directly to the retina using a tack (Fig. 3.3). The microelectrode array is connected to the external system via a flexible cable. When used in conjunction with a small video camera (usually mounted on a pair of glasses) and an external processing system, it provides a rudimentary form of sight to the implanted subjects [132, 133, 134]. The next generation of the device, the Argus II, will contain 60 electrodes which will provide the subject with much higher resolution images. The Argus II entered clinical trials at four hospitals in the U.S. in Jan 2007 and was expanded to three European sites (London, Switzerland, and France) in early 2008. At the time of this writing, Second Sight is the only manufacturer with an actively powered permanently implantable retinal prosthesis under clinical study in the United States. The Argus II is the highest electrode count epi-retinal device being made anywhere in the world.

Intelligent Medical Implants[6] of Switzerland is developing a similar device called the Learning Retina Implant System. The device is fabricated on a flexible polymer substrate and is connected to a wireless telemetry/power module. The external receiver/transmitter and camera is mounted on the earpiece of a pair of glasses. The first chronic study of this device began in the fourth quarter of 2005. Results included the identification of simple patterns and the location of light sources. The second chronic

---

[3]http://www.eye-chip.com/
[4]http://www.2-sight.com/
[5]Epi-Retinal implants sit on the surface of the retina and stimulate the underlying cells.
[6]http://www.intmedimplants.de/

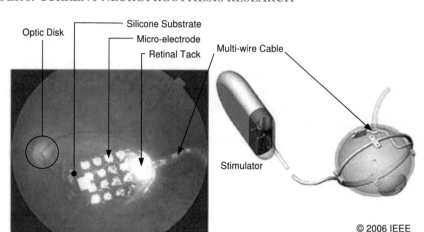

Figure 3.3: The Argus I epi-retinal implant. Left, a photomicrograph of the Argus array *in-situ*. Right, illustration of the components of the Argus system showing the external stimulation unit, multi-wire cable and implanted components [135]. *Reprinted with permission of IEEE EMBS.*

study is on-going and involves the use of the full unit with its wearable camera and pocket sized processing unit.

3.2   COCHLEAR (HEARING) IMPLANTS

Cochlear implant development began in the 1950's when researchers demonstrated that stimulating the auditory nerve could produce auditory sensations. In the sixties scientists learned that stimulation of specific nerve fibers in the auditory nerve via the cochlea was needed to reproduce sound. During this era, three patients were implanted with experimental devices with positive results. Research continued to rapidly advance during the period of the 1970's through the 1980's. In December 1984, cochlear implants were granted FDA approval for nonexperimental implantation into adults. Now systems are FDA approved, safe, and effective. Over 60,000 recipients including children and adults have received these devices to have their hearing at least partially restored.

There is considerable debate in the deaf community regarding the ethics of cochlear implants. The primary argument centers around the deaf communities objection to the mainstream view that deafness is a disability and that deaf people need to be "fixed" through cochlear implants or other means. They argue that there is little or no understanding in the general population that deaf individuals can lead successful, well-adjusted lives without such implants. The counter argument is that while deaf individuals can lead successful lives within the deaf community, it can be very challenging when the individual tries to leave that community and live in the world at large. There have been many tools developed over the years to assist the deaf and hard of hearing: Braille, closed captioning, text-telephones, telecommunications relay services, hearing aids, etc. The cochlear implant is the newest of these tools but is fundamentally different because it is invasive and actually attempts to restore hearing itself rather than providing a "work-around." However, it should be noted that the current technology does not restore hearing per-se,

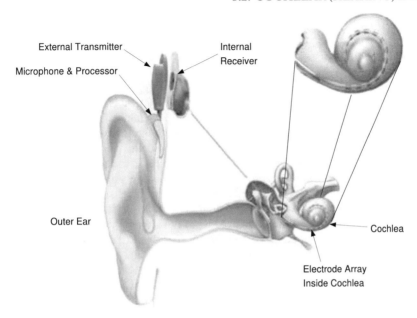

External Transmitter

Microphone & Processor

Internal Receiver

Outer Ear

Cochlea

Electrode Array Inside Cochlea

Figure 3.4: Illustration showing the key parts of a cochlear implant and its placement in the inner ear. *Adapted from an illustration by NIH Medical Arts and provided by the NIDCD [136].*

rather a hearing-like sensation described as an electrical buzzing. A further ethical complication is that the implants work best when they are implanted in the very young, ideally around two years of age before language acquisition has begun. This means that the decision to implant is in the hands of the parents who are likely not deaf and have little or no exposure to the deaf community. Thus, the decision to implant is often made without an understanding of what the deaf community has to offer and with probable bias against any option other than the attempt to restore the child's hearing. Like sign language and braille, it requires extensive training to learn to interpret the information from the implant in a useful way and not all individuals are equally successful at this task. Regardless of which side of the ethical argument you find yourself on, cochlear implant technology continues to evolve and will (probably) be able to restore full, "natural" hearing at some point in the not-too-distant future. Despite its controversy and, for the moment, its shortcomings it still provides another tool in the arsenal to help level the playing field between those who can hear and those who cannot. The National Association for the Deaf has conditionally endorsed cochlear implants as one choice among many choices of assistive tehcnologies for the deaf and hard of hearing.

3.3   COGNITIVE-BASED NEURAL PROSTHETICS: THE RP-2009 INITIATIVE

In the previous chapter, I discussed some of the neural interface technologies used to connect neural prostheses to the nervous system. Cognitive-based neural prosthetics are experimental systems designed to utilize the electrical activity in the brain (via a neural interface) for the purpose of controlling a prosthesis. There are a number of systems under development whose goal is to restore function to amputees, paraplegics, and quadraplegics. One of the most advanced (that I know of) and certainly one of the most ambitious, is DARPA's Revolutionizing Prosthetics (RP)—2009 initiative.

Limb prosthetics have been around probably as long a humans, from simple wooden arms, hooks, or peg-legs to the next-generation device discussed below. Advances in materials, robotics, neural interfaces, and computers have enabled the development of natural looking prosthetics with functionality that approaches that of the natural limb. At DARPATech 2005 (the Defense Advanced Research Projects Administration's annual research meeting), Col. Geoff Ling, MD, outlined his vision to restore function to combat upper limb amputees. He stated that: "This is not a far-off dream... In 4 years, we anticipate having a prosthetic arm that will be controlled identically to the way that we control our biological arms." This large, collaborative program is being carried out at over 30 sites throughout the U.S., Canada, and Europe including the Johns Hopkins University Applied Physics Laboratory (APL) in Laurel, Maryland, which is the project lead. The goal of the project is to have a completed, functioning prototype of a neural-interfaced, prosthetic arm by the end of 2009. Dr. Brett Giroir, Director of DARPA's Defense Sciences

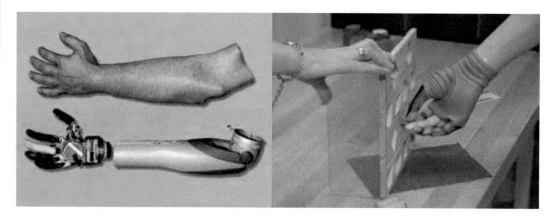

Figure 3.5: DARPA's RP-2009 initiative promises to revolutionize prosthetic limbs for soldiers returning from war; Left: Prototype I of the DARPA arm (below) shown with cosmesis (above); Right: Demonstration of hand dexterity by one of the patients in the study. *Left photo courtesy of the Rehabilitation Institute of Chicago and the Johns Hopkins University APL; Right photo courtesy of the Johns Hopkins University APL.*

Office, stated at the DARPA 25th Systems and Technology Symposium in 2007 that, "...if warfighters could play the piano before their amputation, they would again play the piano with their new prostheses." While Giroir's vision may seem more science fiction than fact, the RP-2009 team is turning it into reality.

The term "degrees of freedom" or DOF when applied to artificial limbs refers to how many ways the limb can move (a simple hinge has 1 degree of freedom, a universal joint has 2). While the degree of movement (DOM) is the number of axes on which movement can occur relative to a reference plane. Thus, a limb may have a higher number for the DOF than the DOM since some of the movements (relative to a given reference plane) permitted by the joints are redundant. For example, how many different ways can you position your arm which result in the fingers of your hand touching your shoulder? Because of its high number of DOF, the joints of the arm permit several possibilities of positioning to move the hand along the same axis of motion.

The prototype-1 system utilizes a Cobot drive [137, 138] located in the forearm to control the movements of the fingers, hand, and wrist. It provides 25+ degrees of freedom and 18 degrees of movement, making it a very good approximation of the natural limb. This prototype provided a testbed for many aspects of the design, construction, testing, and actual use of the arm using control signals derived from muscles. The first prototype did not incorporate direct neural control and sensory feedback. Instead, it is controlled using skin electrodes placed over the pectoral muscle on the same side of the body as the prosthesis. Following the loss of the natural limb the amputee undergoes a surgical procedure which sub-divides the pectoral muscle. The nerves that once served the arm are re-directed into the newly created subdivisions. Activation (contraction) of different parts of the muscle control different movements of the prosthesis. After some training the amputee is able to use the prosthesis much as they would their natural limb. Unfortunately, this method of control is not adequate for the prototype-2 system which will have more degrees of movement than the first prototype.

The prototype-2 system has 22 degrees of movement and incorporates over 80 tactile elements for sensing pressure, vibration, heat flow, and proprioception. The prosthesis will be connected to the users nervous system via either a nerve implant or a brain implant. The hand is capable of producing a 9 kg pinch and a 32 kg grasp which is more than most people can manage with their natural limb. The neural interface component is being developed at the University of Utah and will be the "hair brush" type of array discussed earlier. These arrays have been used in humans as part of several other experimental systems. The Utah slant array technology will be used to create an interface to one of the large nerves carrying signals to the arm. After training, nerve activity corresponding to arm movement would be translated into control signals to guide the prosthesis or a computer generated limb in a virtual environment. The real advantage of the neural interface, versus using surface electrodes, will be that the system can stimulate nerve fibers to create the sensation of touch, hot, cold, or even pain in response to the limb's tactile sensors. A direct prosthesis-to-brain interface using the Utah electrode array would be utilized when a suitable nerve is not accessible or there is damage to the spinal cord. In this case, the brain would be directly connected to the computer system controlling the prosthesis thus bypassing the damaged or nonfunctional area.

3.4   CARDIAC PACEMAKERS

Cardiac pacemakers have been in use for almost 50 years and have helped countless people, but are often overlooked in the discussion of neuroprosthetics. Defibrillators, also known as implantable cardioverter-defibrillators or ICDs, are not only capable of pacing the heart[7] but can intervene in cases of abnormal heart rhythm. When an abnormal heart rhythm occurs the heart can no longer effectively pump blood and may result in death. Defibrillators can deliver an electric shock allowing the heart to hopefully resume a normal rhythm. This neural-prosthetic technology has had a profound impact on human lives.

---

[7]Pacing refers to the process of artificially regulating the rate and intensity of the heart's contractions.

CHAPTER 4

Conclusion

"The only way of discovering the limits of the possible is to venture a little way past them into the impossible."

- Author C. Clarke -

Before diving into the the ethical questions of the technology itself, I would like to say a few words about how bio-technology makes it's way from the drawing board to clinical application in humans. When a device that is destined for use in humans is developed it is required to undergo extensive testing for safety, tolerability, efficacy, and feasibility. Safety testing a material destined for implantation in the human body often begins by placing it in a cell culture. If the culture dies or becomes abnormal in some way then the material is probably not bio-compatible. Next, a device might be fabricated from the material then implanted in a living specimen. While an animal that is low on the phylogenetic tree (e.g., a crustacean) can sometimes be used, work that is destined for use in humans will often begin in a mammalian species. Rats and mice are commonly used for experiments because of their low cost of procurement and maintenance. Also, there are a number of breeds which exhibit disorders similar to those found in humans making them a good testbed for experimental therapies. Usually extensive testing is done in rats or mice prior to moving to a higher mammal or directly to humans. The use of non-human primates (NHPs) in research is highly controversial, they are expensive to procure, and their cost of maintenance is high. In general, if a device is safe enough to use in a NHP, chances are that it is safe enough to begin limited use in humans. However, it is important to understand that this is not always the case and it is still necessary in certain instances to use NHPs for testing. In these cases there are very strict guidelines regulating the care and use of these intelligent animals. These guidelines dictate practices that are very similar to those required for human patients.

Often there are several cycles of design-test-modify-test before obtaining a prototype device which is a candidate for use in humans. During the prototyping period implanted devices are tested in animals. Researchers work with their institution's Internal Review Board (IRB) which oversees the use of animals (and humans) in research and insures that researchers follow established ethical practices and humane use guidelines. When the prototyping period is complete and the device is deemed ready for human testing, the next step may be to enter pre-clinical trials with a limited number (usually less than 10) of human subjects or apply to the country's regulatory body[1] to begin limited experimental use in humans. If approved, the devices are implanted in a small number ($\sim$20–80) of the most needy patients. For example, experimental cancer drugs are often tested only in patients with terminal prognoses for whom conventional treatments have failed or cannot be used. The primary aims of early testing are usually limited to assessing the safety and tolerability of the device. This type of clinical testing of a treatment on a small group of patients to determine safety and side effects is referred to as a *Phase I* clinical trial[2].

Following successful outcome of Phase I trials the device may be tested on a larger number of patients ($\sim$100–300) to further evaluate its safety and determine its effectiveness. This is referred to as *Phase II* clinical trials. *Phase III* testing is performed in potentially thousands of patients to further assess

[1] E.g., The Food and Drug Administration in the United States.
[2] http://www.clinicaltrials.gov/ct2/info/understand

effectiveness, side effects, and outcomes across a large patient population. Finally, *Phase IV* trials use post-marketing studies to evaluate the device's risks, benefits, and additional uses that were not part of the original study. At each step of the testing process of the device information is collected which will guide designers in making refinements and clinicians in its optimal use. There is also a fast-track approval process for treatments which demonstrate safety and high efficacy. The process of developing a device for use in humans takes years of work and the cooperative efforts of a large number of professionals and patients alike. Bringing a device to market represents a huge investment both in terms of taxpayer dollars in grants that funded the basic science and in terms of money spent by companies to fund clinical testing and product development.

Ideally, we would like to have a highly detailed computer simulation of the human body which could be tweaked many different ways to test the devices under different use scenarios, individual physiological variations, etc. Unfortunately, computers are not nearly powerful enough, our understanding of the body woefully inadequate and there does not exist a modelling language or simulation environment that could even remotely handle the complexity. As technology and our knowledge of the body's systems advances this will change. There are fledgling efforts underway to simulate various body systems in software. Such simulations help us to validate our present knowledge (when the simulation performs as the natural system would), show us our errors and underscore our lack of understanding. I have no doubt that some day a large part of the safety testing of drugs and implantable materials/devices will be performed on computers. Unfortunately, the present state-of-the-art leaves us with the imperfect and often ethically complicated alternative of using intact, animal systems that closely resemble our own for the early stages of testing.

From a financial standpoint, animal testing is sub-optimal since animals do not respond to drugs or implanted materials exactly as humans would. Companies sink billions of dollars into the development of new drugs only to find out in clinical trials that what seemed to work in animals either doesn't work in humans or produces undesirable side-effects. From a materials standpoint, animals heal better, have stronger resistance to infection, and tolerate implants better than humans. So, if an implant is not tolerated well in the animal it will almost certainly not be tolerated in the human. However, just because it is well tolerated in the animal does not necessarily mean the same for the human patient. There has been tremendous change in the past 50 years in the way animals are used in research thanks to various efforts to increase awareness and advance the guidelines for their ethical use. Hopefully, with the continuing advances in cell culture, computers, and the fledgling field of organ culture, we will gradually be able to replace the use of animals with engineered human tissue. Assessing drugs and devices in human tissues grown in the lab should prove far more useful and cost effective. Not that ethical choices should be reduced to financial terms but for large companies what is best for the bottom line is definitely a factor in the decision.

I believe that the ethical use of animals in research is necessary at this point in time, but only insofar as alternatives are not available. I feel that there should be a concerted effort to develop systems which provide a better testbed for drugs and devices than what is currently available using animal models. In my opinion, the best place for animals is in their natural habitat, not caged in the lab or zoo. While there are certain organizations and individuals who believe that *any* use of animals is unacceptable and the research which uses them should not be sanctioned, one must consider the alternatives. If we were to halt the use of animals in research we would be left with testing using cell culture. While cell culture can give us a rough idea regarding biocompatibility of implanted materials, it cannot test immune responses, complex tissue reactions, glial scaring and a host of other things that can happen in the intact animal. Thus, we would have one of two choices:

1. Enter clinical trials with minimally tested drugs and devices, using humans as the initial test subjects.

2. Halt research into the development of new therapies.

In my opinion both of these scenarios are ethically unacceptable from a human standpoint. The use of humans as guinea pigs is not only unethical but illegal and the cessation of research is unethical from the standpoint of alleviating human suffering. I am reminded of a scenario posed in class by one of my ethics professors. It goes something like this:

> You are stranded in a raft adrift on the ocean. Among your number are yourself, two other humans, and a pet chimpanzee. You have water which you collect when it rains, but no food and no way to obtain food from the sea. You have been stranded now for nearly two weeks and you are all on the brink of starvation. It is believed that you will be rescued within the next week if you can just hold out a little longer. *What would you do to survive?*

I was surprised by how much debate this scenario elicited (an entire 50 minute class period as I recall) over what seemed to be a relatively straightforward survival decision. So, I posed the same scenario to my 4 year old son and 7 year daughter who arrived at the response, "eat the monkey" in 15 seconds and 2 minutes, respectively. The debate over the use of animals in research, though analogous, is not quite so cut-and-dried. While I understand the viewpoint of animal-rights activists and respect the strength of their convictions, I cannot in good conscience agree with all of their tenets. I feel that the goal of alleviating human suffering trumps the rights of animals. So, for the time being, we have three alternatives: to use humans as the proverbial "guinea pigs," to stop the development of new therapies, or to use animals in our research. In other words, it's the choice of the lesser of three evils and I choose the latter.

4.1   NEURAL INTERFACING ETHICS

As mentioned earlier, a nervous system provides organisms with the unparalleled ability to adapt to their dynamic environments; something that computers are a long way from achieving. This will almost certainly change given enough time and the continued progress of technology. For the time being, we are very far from the day when human-like machines will be able to match the adaptability and creativity of their creators. Some might say that the best way to achieve this end is the merger of humans with the machines they might create, to others this is a horrific notion that is best left on the pages of science fiction.

While restoration of function is a noble, high-priority research goal, one cannot ignore the potential of this technology for another more far-reaching application, the augmentation of human abilities. In this context the realization of a complete human-machine neural interface could potentially be one of the most important events in human evolution since our distant ancestors evolved opposable digits. Neural prosthetics could not only allow the crippled to walk, the blind to see, and the deaf to hear, but could eventually augment/expand the human mind in ways we can only begin to imagine. Direct Neural Interface (DNI) augmented intelligence would permit users to instantly tap into global databases containing the store of human knowledge. Abstract ideas and thoughts for which language is an inadequate medium could be communicated directly, mind to mind, through a wireless link. The precise number-crunching abilities of computers could be merged with the creative ability of the human mind permitting users to perform complex calculations "in their head." An architect could, for example, imagine a new design for a bridge while simultaneously calculating load factors, material costs, etc. A musician could imagine a musical composition and then through the use of computer interfaced instruments have it performed. Technical ability and manual dexterity would no longer limit the ability of a person to bring to fruition whatever their mind can imagine. Computer-augmented senses could allow scientists and engineers to directly experience with their own virtual eyes any portion of the electromagnetic spectrum for which a sensing modality exists. Astronomers could peer into the furthest reaches of the universe and doctors could look into patients using their own cybernetically enhanced vision.

Immersive virtual reality (IVR) refers to totally computer generated environments in which the user has complete motor control and sensory feedback just as in the real world. The current technologies of body position tracking devices, headgear, haptic feedback, etc., which attempt to approximate IVR are as primitive as stone knives when compared to DNI. Neural interface based IVR could put the user in a virtual environment that seems as real as the physical world. One of the many promises of DNI-IVR is to bring people living in physically distant locations together in virtual environments where they can interact with the same richness of experience as a face to face meeting. Thus, a person living in the Australian outback could instantly be in the heart of downtown "Cyberopolis" where they could interact with millions of other connected users from around the world. Likewise, virtual schools and colleges could bridge the education gap between rural and urban areas. People might go to work by "plugging in" and joining their co-workers in a virtual office.

The word "avatar" is taken from Hindu mythology and refers to the reincarnation of an advanced or enlightened soul on earth. In virtual reality parlance an avatar is a 3D construct which is used to represent a user in a virtual environment and may or may not look anything like the user it serves. When a robot is used as an avatar it becomes a tool for immersive tele-presence. Through the use of DNI-IVR and an avatar a person could be instantly anywhere in the real or virtual world. For example, a business person in Nome, Alaska would rent an android avatar in Beijing, China. Then, using DNI-IVR, have his or her sensorium mapped onto the body of the avatar in order to conduct business in the real world far away from their physical location. Likewise, an astronaut could explore another world without ever leaving the safety of earth. Others who want to share the experience could participate as observers, feeling everything the astronaut feels but not controlling the avatar. Recordings of the event could be stored allowing future generations to directly re-experience it in a first person sense. Other events could likewise be recorded allowing people to later experience history through the eyes and minds of the people who were there.

A complete interface to the brain might permit a user to copy part of their knowledge/personality into an avatar which then could act as an agent working on the user's behalf to perform some duty. While performing its task the avatar would maintain communication with the user and exchange information as needed. This is analogous to a computer program forking processes then communicating with those processes through a pipe[3]. Using a similar mechanism, and prior to the expiration of a user's physical body, their knowledge might be uploaded into a global repository. This global repository might exist as a type of generalized expert system or "counsel of elders" composed of millions of incorporeal intelligences which could be consulted whenever their particular expertise is needed. Future scientific advances might allow a new body to be grown for an incorporeal intelligence which could then be downloaded into it to become corporeal again.

The use of neuroprosthetics beyond what is strictly required for rehabilitation is fraught with both practical and ethical concerns. To use a more familiar example, cosmetic surgery is widely available to anyone with the motivation and financial resources. One can safely and easily be shaped and enhanced to look younger, slimmer, or better endowed. With a bit more effort and financial commitment, one can even change their sex or become something in between male and female. All of this has come about in the last 50 years (give-or-take) with the advance in surgical techniques, implantable materials and varying degrees of social acceptance. What will the next 50 bring? Aside from the regulation of medical standards

---

[3]A communication channel used to exchange data.

of practice, we do not prohibit people from altering their bodies. We would not say to an actor or actress "you can't have any more cosmetic surgery because you look too good and are taking jobs from other less attractive professionals." Likewise, what would be the rationale for limiting and/or preventing people from enhancing themselves with safe and effective cybernetic implants should they one day be available?

Sci-fi author and father of the "Cyberpunk" genre, William Gibson, wrote extensively on a fictitious, future society whose denizens often spent a great deal of their time connected to the vast virtual world he termed "Cyberspace." As with many things which promise great benefits there is the potential for abuse. The dark side of DNI-IVR is that it could become a "drug" more powerful and addictive than any present day pharmaceutical. Persons looking for an escape could become lost in a world of anything-goes synthetic reality that is seemingly as real as the physical world. A whole new class of mental disorders could arise from people who spend too much time in IVR environments, who are unable to cope with what is presented in those environments, and/or lose the ability to distinguish them from the "real" world.

A user plugged into a DNI which completely disconnects them from all the physical sensation of their real body could die of thirst or hunger if nutrition is not provided. Furthermore, our sensory and motor systems don't respond on nanosecond time scales. When we first learn to grasp an object, crawl, and eventually walk, our brains have learned all these tasks while compensating for built in delays, errors, or irregularities unique to a person's own natural system. A user's brain directly connected to a virtual environment could experience instantaneous feedback without the delays imposed by their natural sensory/motor systems. New forms of physical disability could result from the disparity between the responsiveness of virtual limbs and the users own limbs. A user who spends too much time in an ultra-responsive virtual world could find their coordination, motor control, and sensory perception severely affected when they are "unplugged." A person could literally forget how to use their own limbs or process their own natural sensory input.

The ultimate neural interface would seamlessly integrate cybernetic and biological components. With advances in the fields of molecular biology, genetics, and nanotechnology, new types of interfacing paradigms may become possible. For example, presently there is an extensive effort to create materials which can be implanted in the body without fear of adverse or undesirable reactions. In the future it may be possible to genetically modify a person's body to more readily accept implants. Alternately, cultured implants containing both electronic and neuronal components would be grown in labs, then placed in the brain or body where they continue to grow and make connections. Such an implant might be custom made for a specific individual with the biological components created from their own cells or DNA. Further genetic modification could enhance strength, intelligence, or possibly re-engineer the brain to make it physically easier to access for cybernetic implantation. Might a human so radically modified by genetic alterations and implants be considered a new species, perhaps even a next step in our evolution? What would this mean for the future of the human race, particularly those who have not been modified?

Thus far we have discussed human enhancement using neuroprosthetics, but what of using biological elements to enhance computers? Before continuing I should point out that there many technical hurdles to overcome with using living neural networks as components in computers. However, based upon some of the work discussed earlier, in the *distant* future it may become possible to build arrays of biological processing elements containing specially grown neural networks. These networks could be incorporated into electronic/photonic systems, endowing them with more human-like abilities. These biological components could provide the essential front-end for natural language recognition, speech synthesis, vision processing, and other functions where humans and animals presently outperform computers. It is possible that the next paradigm shift in computer technology will be "wet." That is, computer systems of the future might consist of organic and in-organic components with the organic components sustained by a nutrient and oxygen-rich solution not unlike blood. Since biological systems are (at present) far smaller and more portable than computers, organo-synthetic processors could become the technology that paves the way for the "android" so often written about in science fiction. That is, a true synthetic being comprised of biological and technological components; having blood, needing air to breathe, and consuming nutrients to power its systems.

This brings to mind the larger ethical question which is, "would such systems have rights?" Would they be subject to the same treatment before the law as humans? If not, what would be the rationale for denying them these protections? If one were to use the "sentience" argument then how do we judge sentience? If the argument simply is, "machines don't have rights," then aren't we just machines of a different type? We, having evolved to our present state through natural processes, instead of being built by another race of beings. Fortunately for society and jurisprudence, such systems are unlikely to exist even in our great grandchildren's lifetimes, but the ethical dilemma posed by the emergence of engineered beings makes for lively academic debate as well as interesting fiction.

Science is many years away from producing the kind of technologies discussed above (Tab. 4.1), however, given enough time and effort it is easy to imagine a day when such things will become possible. There are those like Dr. José Delgado who believe the type of neural interface hoped for by the trans-humanists[4] and feared by the mind-control, conspiracy theorists is simply not realizable [139]. Others believe that producing technologies such as hybrid silicon-neuron computer components is so fraught with technical challenges that it may never be practical. While few would argue against using neuroprosthetics to help the disabled, what about the question of enhancing otherwise normal individuals? Banning particular uses of cybernetic enhancement or lines or research through legislation will prove no more effective than bans on cloning or stem cell research. Someone, somewhere in the world will conduct the research, develop the product and perform the surgery because there is money to be made. In these situations the law of supply and demand has more effect than the laws of some body of government.

The difficulty is that once the body of knowledge pertaining to a particular technology becomes known it cannot be withdrawn again from public knowledge. Once that knowledge is realized in a deliverable therapy, device, or treatment no one government will be able to regulate its use. If you can't obtain the desired enhancement in the U.S. then one could simply travel to Thailand, China, Europe or elsewhere. At that point the ethical discussion will be rendered largely academic and we will find ourselves dealing with the ramifications of a two tier society, those who are enhanced and those who are not. The proper way to address the ethical issues of neural interfacing is to begin discussion well ahead of the availability of such technologies, bringing as many scientists, business people, and countries to the table as possible. A strong international regulatory structure may help insure the safety of therapies that are developed around neural interfacing technologies and moderate potential deleterious social effects of wide spread non-therapeutic use.

The ethical and social ramifications of producing a society of cybernetically or genetically enhanced individuals should not be ignored. Almost certainly these types of enhancements will be expensive (especially at first) and as such only available to a wealthy few. Employer sponsored cybernetic enhancement could become both a benefit and condition of employment. Such "benefits" could also become a means for unscrupulous employers to make it difficult for employees to quit; especially if those enhancements are proprietary and not easily undone. Normal, unenhanced individuals who cannot afford, or do not want to be upgraded could find themselves relegated to menial jobs; becoming the untouchables of a technologically imposed caste system.

The formation of a total, direct neural interface between humans and machines holds both great promise and potentially dire consequences. However, neural interfacing is one of a few technologies that has the potential to completely change the face of human society by bridging the language, cultural, edu-

---

[4]http://www.transhumanism.org

Table 4.1: Some examples of neural-based technologies discussed in this book and "guesses" as to when they might be fully realized (assuming continuing research efforts)

	CA	CT	NF	FF	?
Auditory Neuroprosthesis (Ear)			•		
Whole-Eye Neuroprosthesis				•	
† Artificial Retina		•			
Olfactory Neuroprosthesis (Nose)			•		
‡ Cochlear Implant	•				
† EEG Computer Interface	•				
† DNI Computer Interface		•			
DNI Based Remote Telepresence			•		
Total DNI based Immersive VR				•	
Mind Reading (EEG or some other technology)					•
Mind/Memory Uploading/Downloading					•
Mind Linking & Cognitive Expansion					•
Hybrid Silicon-neural Computer Components				•	

CA=Commercially Available Now, CT=Presently in Clinical Trials,

NF=Near Future (50-100 yrs), FF=Far Future (100-200 yrs), ?= Very distant future, if ever.

† Indicates a technology in the early stages of development

‡ Indicates an advanced but still evolving technology

cational, and physical barriers that have traditionally divided humans along so many lines. The realization of the technology needed to accomplish a DNI has begun. So as the Cyborg makes its way from the pages of science fiction into the pages of scientific journals it will be up to both present and future generations to decide how humanity takes what could amount to a significant, next step in its evolution.

CHAPTER 5

Resources for Students

This Chapter contains an abbreviated listing of groups conducting neural-interfacing related research. This listing is based on the work discussed in this book and *is not* a comprehensive listing of all the work occurring in the field. Of the work discussed, only the contact information for those researchers known to be conducting neural engineering research at the time of this writing are included. A more complete and up-to-date list of researchers, universities, groups, and companies is available at http://www.neuropunk.org.

If you are contemplating a career in neural engineering there are several things you may want to consider. First, as I have already mentioned, the field is highly interdisciplinary so a broad education is often advantageous. A general knowledge of the gross anatomy, cytoarchitecture, and functional organization of the nervous system (well beyond what is covered in this book) is essential for persons interested in pursuing a career in Neural Engineering. At the very least you should have some familiarity with the organization of the nervous system and what functions its various structures are thought to perform. A good, senior- or graduate-level neuroscience course should fulfill this requirement. Other knowledge/skills you might want to accumulate during your undergraduate or graduate training may include:

- Technical Writing and the ability to illustrate your ideas: an absolute must—you have to be able to communicate your ideas and concepts with other professionals. If drawing is not your thing, learn to use Computer Aided Drafting and Design (CADD) software.

- Computer programming: at the very minimum learn C/C++ or Java, Neuro-prosthetics are run by software, even if you are not the one writing the program a knowledge of computer programming will help you in your interactions with other team members.

- Electronics:
  - A working knowledge of electronics and basic principles such as resistance, impedance, and capacitance. This is often taught as part of college physics.
  - The ability to design and fabricate basic analog circuitry (e.g., an amplifier) and basic digital logic circuitry. If your university does not offer this, check the local community college.

- Biology: Freshman- and sophomore-level possibly including cellular/molecular biology.

- Mathematics through introductory Calculus.

- College-level physics.

Remember any advice is just that, advice. Don't take the opinions of one person (including myself) as the final word on what you should be doing. If I had relied on the advice of my high school guidance counselor I would never had gone to college, earned a Ph.D. and gone on to do other things like write this book. Ultimately, it's your decision and you are the one that has to live with the choices you make. To make infomed decisions about your future you need more than just sage advice, you need the one thing you may not have at this point, experience. Many universities and laboratories offer summer internships for high school/college students and lab rotations for graduate students. Internships and rotations will allow you to explore your interests, get a taste of what working in a lab is really like. More to the point, it will put you in contact with those who can best advise you.

In addition to the subjects mentioned above there are several unrelated but important things you should consider. We live in a global economy where modern communication and transportation are shrinking the distances between countries. It is important that you have some knowledge of cultures other than the one you in which you were raised. The best way to learn about other cultures is through foreign language courses and foreign exchange programs. We as Americans tend to have a myopic view of the world since we live in a large, more-or-less homogeneous country which "officially" recognizes only a single language[1]. The world is a large diverse place filled with many different cultures, languages and peoples. Financially, most of the market capitalization is vested in the rest of the world, not the US. Additionally, there are numerous neural engineering research efforts in Europe, Japan, and China. To give youself the most options for having the career of your choice, commit to learning a second language to the point of fluency.

If possible, work some business management courses into your schedule or take the extra time needed to get a minor (or a second degree) in business. In today's environment graduate students become post-docs then become professors only to discover that they now have to run a lab, manage personnel, obtain funding or even spin off businesses from their research. While at the same time trying to publish their work and navigate through a university's tenure and promotion process. They have little or no experience delegating tasks, managing people, writing business proposals, etc. As a student, you only have to worry about completing your own work. Research professionals not only have their own work but often have to manage a team. A good business management curriculum will provide you with the basic skills you need so you won't be starting "cold" when confronted with managing a lab or a business. Furthermore, if you decide to go into industry instead of academia you will find some business management knowledge invaluable.

Completing a degree is not the end of your studies but the beginning. One purpose of college is to transition you from high school where you are told what assignments to complete, how they are to be completed and when they are due, to working and learning independently. Technology is constantly changing, our knowledge of the brain grows and is revised frequently and new tools are always being developed. We scientists must commit to a lifelong journey of self education. College and graduate school are essential stepping stones along the way designed to give you the foundation you need, but the foundation is only the beginning of any well-made structure.

5.1   CONTACT INFORMATION

Please visit the website listed for the professors/research groups prior to contacting them. Many of the groups have information on their site regarding educational opportunities. When making intital contact by e-mail please check the group's homepage(s) for the current address.

---

[1]Although Spanish is the un-official, second language in the U.S.

5.1.1 JAPAN

Cardiovascular Research Institute
http://www.ncvc.go.jp/english/res/rese.html

Dr. Masaru Sugimachi, M.D.
National Cardiovascular Center Research Institute
Department of Cardiovascular Dynamics
5-7-1 Fujishiro-dai   Phone:+81 06 6833 5012
Suita, Osaka 565-8565   Fax: +81 06 6835 5403

Laboratory for Advanced Brain Signal Processing
http://www.bsp.brain.riken.jp

Dr. Andrzej Cichocki, Ph.D.
RIKEN Brain Science Institute
Laboratory for Advanced Brain Signal Processing
2-1 Hirosawa   Phone:+81 48 467 9668
Wako, Saitama 351-0198   Fax: +81 48 467 9686

5.1.2 EUROPE

Intelligent Data Analysis Group
http://ida.first.fraunhofer.de/homepages/ida/

Prof. Dr. Klaus-Robert Müller, Ph.D.
Kekuléstr. 7
D-12489 Berlin   Phone: +49 0 30 6392 1860
Germany   Fax: +49 0 30 6392 1805

The Fraunhofer Institute for Biomedical Engineering
http://www.ibmt.fraunhofer.de/fhg/ibmt_en/biomedical_engineering/
medical_engineering_neuroprosthetics

Prof. Dr. Klaus-Peter Hoffmann, Ph.D.
Fraunhofer-Institut für Biomedizinische Tech. IBMT
Dept. of Medical Engineering and Neuroprosthethics
Ensheimer Straße 48
66386 St. Ingbert   Phone: +49 6894 980 401
Germany   Fax: +49 6894 980 400

Institute of Microsystem Technology - IMTEK
http://www.imtek.uni-freiburg.de/index_en.php

Prof. Dr. Thomas Stieglitz, Ph.D.
Laboratory for Biomedical Microtechnology
Georges-Koehler-Allee 102
D-79110 Freiburg   Phone: +49 0 761 203 7471
Germany   Fax: +49 0 761 203 7472

Institut de recherche IDIAP
http://www.idiap.ch/current_projects.php?project=27

Centre du Parc   **info@idiap.ch**
Av. des Près-Beudin 20   http://www.idiap.ch/jobs.php
Case Postale 592
CH-1920 Martigny   Phone: +41 27 721 77 11
Switzerland   Fax: +41 27 721 77 12

Max Planck Institute of Biochemistry
http://www.biochem.mpg.de/mnphys/

Prof. Dr. Peter Fromherz, Ph.D.
Max Planck Institute of Biochemistry
Department of Membrane and Neurophysics
Am Klopferspitz 18
D-82152 Martinsried
Germany

Phone: +49 89 8578 2821
Fax: +49 89 8578 2822

Warwick Biosensors Group
http://template.bio.warwick.ac.uk/staff/ndale

Prof. Dr. Nicholas Dale, Ph.D.
University of Warwick
Department of Biological Sciences
Gibbet Hill Road
Coventry CV4 7AL
United Kingdom

Phone: +44 0 24 7652 3729
Fax: +44 0 24 7657 2594

5.1.3   UNITED STATES

The Andersen Lab
http://vis.caltech.edu/

Dr. Richard A. Andersen, Ph.D.
California Institute of Technology
Division of Biology
MC 156-29
Pasadena, CA 91125

Phone: +1 626.395.4951
Fax: +1 626.449.0756

The Applied Physics Laboratory
http://www.jhuapl.edu

Dr. Stuart Harshbarger, Ph.D.
The Johns Hopkins University
Applied Physics Laboratory
11100 Johns Hopkins Rd
Laurel MD 20723-6099

Phone: +1 240 228 5000
Fax: +1 240 228 1093

Biomimetic MicroElectronic Systems Engineering Research Center.
http://bmes-erc.usc.edu

Dr. James D. Weiland, Ph.D.
Biomimetic MicroElectronic Systems Eng. Research Ctr.
1450 San Pablo Street
DVRC 130
Los Angeles, CA 90033-1035

Phone: +1 323 442 6786
Fax: +1 323 442 6790

Biomimetic MicroElectronic Systems Engineering Research Center
http://bmes-erc.usc.edu

Dr. Gerald E. Loeb, M.D.
University of Southern California
Denney Research Building B12, MC1112
1042 Downey Way
Los Angeles, CA 90089

Phone: +1 213 821 1112
Fax: +1 213 821 1120

Brain-Computer Interface Research Project
http://www.bciresearch.org/

Dr. Jonathan R. Wolpaw, M.D., Ph.D.
New York State Department of Health
Wadsworth Center
P.O. Box 509
Albany, New York 12201-0509

http://www.wadsworth.org/docs/education.shtml
Phone: +1 518 473 3631
Fax: +1 518 486 4910

Center for Adaptive Neural Systems
http://ans.asu.edu/

Dr. Ranu Jung, Ph.D.
Dr. Jimmy Abbas, Ph.D.
Arizona State University
Fulton School of Engineering
PO Box 874404
Tempe, AZ 85287-4404

Phone: +1 480 965 9489
Fax: +1 480 727 7624

Center for Microelectrode Technology
http://www.mc.uky.edu/cenmet

Dr. Greg Gerhardt, Ph.D.
University of Kentucky
306 Davis Mills Bldg.
800 Rose Street
Lexington, KY 40536-0098

Phone: +1 859 323 4531
Fax: +1 859 257 5310

Center for Neural Engineering
http://www.neural-prosthesis.com/

Dr. Theodore W. Berger, Ph.D.
Department of Biomedical Engineering
164 Denney Research Building
University of Southern California
Los Angeles, CA 90089-1111

info@neural-prosthesis.com

Phone: +1 213 740 8017
Fax: +1 213 821 2368

Center for Wireless Integrated Microsystems ERC
http://www.wimserc.org

Dr. Kensall D. Wise
The University of Michigan
Dept. of Electrical Eng. and Computer Science
2401 EECS Bldg.
1301 Beal Avenue
Ann Arbor, MI 48109-2122

Phone: +1 734 764 3346
Fax: +1 734 763 9324

Computational NeuroEngineering Laboratory
http://www.cnel.ufl.edu/

Dr. José Carlos Príncipe, Ph.D.
University of Florida
Electrical & Computer Engineering Department
NEB 451, Bldg #33
Gainesville, FL 32611

Phone: +1 352 392 2662
Fax: +1 352 392 0044

The Donoghue Laboratory
http://donoghue.neuro.brown.edu

Dr. John Donoghue, Ph.D.
Brown University
Department of Neuroscience
185 Meeting Street
Providence, RI 02912

Phone: +1 401 863 9524
Fax: +1 401-863-1074

Neural Engineering Center
http://nec.cwru.edu/

Dr. Dominique M. Durand, Ph.D.
Case Western Reserve University
Department of Biomedical Engineering
Wickenden 309
Cleveland, OH 44106

Phone: +1 216 368 3974
Fax: +1 216 368 4872

Neural Engineering Laboratory

http://nelab.engin.umich.edu

Dr. Daryl R. Kipke, Ph.D.
1101 Beal Ave.
Ann Arbor MI 48109-2110

Phone: +1 734 764 3716
Fax: +1 734 647 4834

Neuroprosthetics Research Group

http://nrg.mbi.ufl.edu/

Dr. Justin C. Sanchez, Ph.D.
University of Florida
Department of Pediatrics
P.O. Box 100296, JHMHC
Gainesville, Florida, 32610

Phone: +1 352 846 2180
Fax: +1 352 392 2515

Nicolelis Lab

http://www.nicolelislab.net/NLnet_Load.html

Dr. Miguel A. L. Nicolelis, M.D., Ph.D.
Duke University Medical Center
Department of Neurobiology
Box 3209, DUMC
Durham, NC 27710

Phone: +1 919 684 4580
Fax: +1 919 684 4431

The Pine Lab

http://www.its.caltech.edu/$\sim$pinelab

Dr. Jerome Pine, Ph.D.
California Institute of Technology
Dept. of Physics, Mathematics and Astronomy
MC 256-48
Pasadena, CA 91125

Phone: +1 626 395 6677

Wightman Research Group

http://www.chem.unc.edu/people/faculty/wightmanrm/rmwgroup/

Dr. R. Mark Wightman, Ph.D.
The University of North Carolina at Chapel Hill
Department of Chemistry
Campus Box 3290
Chapel Hill, NC 27599-3290

Phone: +1 919 962 1472
Fax: +1 919 962 2388

5.2   LINKS TO FREE RESOURCES ON THE INTERNET

Table 5.1:	
Location	**Name and URL**
US	**NINDS - National Institute for Neurological Disorders and Stroke** Neural Prosthesis Program, `http://www.ninds.nih.gov/funding/research/npp/index.htm`
US	**National Science Foundation** `http://www.nsf.gov/funding/pgm_summ.jsp?pims_id=5147\&from=fund`
US	**Neuropunk.Org - Online Encyclopedia of Neural Engineering** `http://www.neuropunk.org`
DE	**BCI-Info - International Portal for Brain Computer Interfaces** `http://www.bci-info.tugraz.at`
US	**Scholarpedia.Org - The free peer reviewed encyclopedia** `http://www.scholarpedia.org`

5.3   FURTHER READING

- *From Neuron to Brain*, John G. Nicholls et al., ISBN: 0-87893-439-1.

- *Brain-Machine Interface Engineering*, by José Principe and Justin Sanchez, ISBN: 9781559820349.

- *Toward Replacement Parts for the Brain*, by Theodore Berger and Dennis Glanzman, ISBN: 0262025779.

- *Cybernetics: or Control an Communication in the Animal and Machine*, by Norbert Wiener, ISBN: 026273009-X.

- *Electrochemical Methods: Fundamentals and Applications*, by Allen J. Bard and Larry R. Faulkner, ISBN: 0-47104-372-9.

ACKNOWLEDGMENTS

I would like to thank the following individuals without whom this book would not have been possible. My wife for her patience and support. Fred Prior, Linda Larson-Prior, and Greg Gerhardt for their guidance and mentoring. Jerry Pine and Peter Huettl for their feedback. Last, but not least, a special thanks to all of those who contributed graphics material.

ABOUT THE AUTHOR

A native of Virginia, Dr. Coates received her undergraduate degree in biology from Virginia Commonwealth University in 1995, her Ph.D. in Neuroscience from the Pennsylvania State University in

2001 [27], and completed postdoctoral work at the University of Kentucky's Center for Microelectrode Technology. She is presently a private research consultant who specializes in issues surrounding neu-ral interface construction. Her Company (TC Design Group, LLC) is the sponsor of Neuropunk.Org (http://www.neuropunk.org), a nonprofit educational project dedicated to the free dissemination of quality information relating to neural engineering. When not building Cyborgs she enjoys exploring planet Earth, SCUBA diving, and pursuing her interests in music and art.

Commentary and discussion on this book can be found at:
https://susannecoates.com/neural-interfacing

AddressforCorrespondence
Susanne D. Coates,
PO Box 554
Frederick MD 21705-0554

This book was written entirely using open-source tools running on the Linux operating system.

Bibliography

[1] A. Abbott. In search of the sixth sense. *Nature*, 442(13):125–127, Jul. 2006. DOI: 10.1038/442125a

[2] J. G. Nicholls, R. A. Martin, B. G. Wallace, and P. A. Fuchs. *From Neuron to Brain*, vol. 1. Sinauer Associates, Inc., Sunderland, MA, 4th ed., 2001.

[3] K. H. Pribram. *Brain and Perception: Holonomy and Structure in Figural Processing*. John M Maceachran Memorial Lecture Series. Lawrence Erlbaum, Mahwah, NJ, 1st ed., 1991.

[4] K. H. Pribram. *Rethinking Neural Networks: Quantum Fields and Biological Data*. Lawrence Erlbaum, Mahwah, NJ, 1st ed., 1993.

[5] J. R. Wolpaw, N. Birbaumer, D. J. McFarland, G. Pfurtscheller, and T. M. Vaughan. Brain-computer interfaces for communcation and control. *Clin. Neurophysiol.*, 113:767–791, 2002. DOI: 10.1016/S1388-2457(02)00057-3

[6] The stanford encyclopedia of philosophy. World Wide Web.

[7] H. Gray. *Anatomy of the Human Body*. Henry Gray (1825 - 1861). Lea and Febiger, Philadelphia, 20th ed., 1918.

[8] S. Finger. *Origins of Neuroscience: A History of Explorations into Brain Function*. Oxford University Press, 1st ed., Oct. 11, 2001.

[9] G. Perea and A. Araque. Synaptic information processing by astrocytes. *J. Physiol. Paris*, 99:92–97, 2006. DOI: 10.1016/j.jphysparis.2005.12.003

[10] E. R. Kandel, J. H. Schwartz, and T. M. Jessell. *Principles of Neural Science*. McGraw-Hill Medical, New York, 4th ed., Jan. 2000.

[11] H. B. Barlow. *Sensory Communication*, Chapter 13, pp. 217–234. The MIT Press, Cambridge, MA, 1st ed., 1961. Possible principles underlying the transformations of sensory messages.

[12] F. Rieke, D. Warland, R. de Ruyter van Steveninck, and W. Bialey. *Spikes: Exploring the Neural Code*. The MIT Press, Cambridge, MA, 1st ed., 1997.

[13] E. D. Adrian and Y. Zotterman. The impulses produced by sensory nerve endings, part ii: The response of a single end organ. *J. Physiol.*, 61(2):151–171, Apr. 23, 1926.

[14] D. O. Hebb. *The Organization of Behavior: A Neuropsychological Theory*. John Wiley & Sons Ltd., New York, 1st ed., 1949.

[15] G. J. Stuart and B. Sakmann. Active propagation of somatic action potentials into neocortical pyramidal cell dendrites. *Nature*, 6:69–72, Jan. 1994. DOI: 10.1038/367069a0

[16] H. Markham, M. H. Lübke, M. Frotscher, and B. Sakmann. Regulation of synaptic effi-cacy by coincidence of postsynaptic aps and epsps. *Science*, 275(5297):213–215, Jan. 10, 1997. DOI: 10.1126/science.275.5297.213

[17] G. J. Stuart, S. G. Spruston, B. Sakmann, and M. Häusser. Action potential initiation and backpropagation in neurons of the mammalian cns. *Trends Neurosci.*, 20(3):125–131, Mar. 1997. DOI: 10.1016/S0166-2236(96)10075-8

[18] I. R. Winship and T. H. Murphy. In vivo calcium imaging reveals functional rewiring of single somatosensory neurons after stroke. *J. Neurosci.*, 28:6592–6606, Jun. 2008. DOI: 10.1523/JNEUROSCI.0622-08.2008

[19] T. J. Sejnowski and C. R. Rosenberg. Parallel networks that learn to pronounce english text. *Complex Syst.*, 1:145–168, 1987.

[20] H. Wang. *Reflections on Kurt Gödel*. MIT Press, Cambridge, MA, 1st ed., Mar. 1990.

[21] S. M. Ulam. Reflections on the brain's attempts to understand itself. *Los Alamos Science*, pp. 283–287, Oct. 1987.

[22] H. Moravec. *Robot: Mere Machine to Trancendent Mind*, vol. 1, Oxford University Press, New York, 1st ed., 1999.

[23] C. A. Skarda and W. J. Freeman. How brains make chaos to make sense of the world. *Behav. and Brain Sci.*, 10:161–195, 1987.

[24] R. Linsker. Local synaptic learning rules suffice to maximize mutual information in a linear net-work. *Neural Computat.*, 4:691–702, 1992. DOI: 10.1162/neco.1992.4.5.691

[25] A. J. Bell and T. J. Sejnowski. An information-maximization approach to blind separation and blind deconvolution. *Neural Computat.*, 7:1129–1159, 1995. DOI: 10.1162/neco.1995.7.6.1129

[26] C. E. Shannon. A mathematical theory of communication. *Bell Syst. Tech. J.*, 27:379–423, 623–656, Jul. - Oct. 1948. DOI: 10.1145/584091.584093

[27] S. D. Coates. *A Neural Network Based Cybernetic Interface for Identification of Simple Stimuli Based on Electrical Activity in an Intact Whole Nerve*. Ph.D. thesis, Pennsylvania State University, 2001.

[28] P. Pajunen, A. Hyvärinen, and J. Karhunen. Nonlinear blind source separation by self-organizing maps. In *Proc. of the 1996 International Conference on Neural Information Processing (ICONIP'96)*, vol. 2, pp. 1207–1210. Springer-Verlang, Hong-Kong, 1996.

[29] J. Karhunen, E. Oja, L. Wang, R. Vigario, and J. Joutsensalo. A class of neural networks for indepen-dent component analysis. *IEEE Trans. Neural Netw.*, 8:486–504, 1997. DOI: 10.1109/72.572090

[30] Y. Tan, J. Wang, and J. M. Zurada. Nonlinear blind source separation using a radial basis function network. *IEEE Trans. Neural Netw.*, 12:124–134, 2001.

[31] C. H. Zheng, D. S. Huang, K. Li, G. Irwin, and Z. L. Sun. MISEP method for postnonlinear blind source separation. *Neural Comput.*, 19:2557–2578, Sept. 2007. DOI: 10.1162/neco.2007.19.9.2557

[32] M. Imai. *Kaizen: The Key To Japan's Competitive Success*. McGraw-Hill/Irwin, 1st ed., Nov. 1, 1986.

[33] N. Wiener. *Cybernetics: or Control and Communication inthe Animal an the Machine*. MIT Press, Cambridge, MA, 2nd ed., 1999.

[34] M. Clynes and N. S. Kline. Cyborgs and space. *Astronautics*, p. 27, Sept. 1960.

[35] D. M. Durand. Editorial by Dominique M. Durand. *J. Neu. Eng.*, 4(4), Dec. 2007.

[36] D. Schmorrow. Augmented cognition: Building cognitively aware computational systems. Technical report, Defense Advanced Research Projects Administration (DARPA), Anahiem, CA, Jul. 30, Aug. 7, 2002.

[37] G. S. Brindley and W. S. Lewin. The sensations produced by electrical stimulation of the visual cortex. *J. Physiol.*, 196:479–493, 1968.

[38] J. M. R. Delgado, V. Mark, W. Sweet, F. Ervin, G. Weiss, G. Bach-y Rita, and R. Hagiwara. Intracerebral radio stimulation and recording in completely free patients. *J. Nerv. Ment. Dis.*, 147:329–340, 1968.

[39] J. M. R. Delgado. *Physical Control of the Mind*, vol. 1. Harper and Row, New York, 1st ed., 1969.

[40] J. M. R. Delgado. Personal communication, May 2008.

[41] O. Foerster. Beitriige zur pathophysiologie der sehbahn und der sehsphare. *J. Psychol. Neurol.*, 39:463–485, 1929.

[42] F. Krause and H. Schum. *Neue deutsche Chirurgie*, vol. 49a, pp. 482–486. Enke Publishers, Stuttgart, Germany, 1931.

[43] G. Holmes. The organization of the visual cortex in man. *Proc. Roy. Soc. London Ser. B (Biol.)*, 132:348–361, 1945. DOI: 10.1098/rspb.1945.0002

[44] G. S. Brindley, P. E. K. Donaldson, M. D. Falconer, and D. N. Rushton. The extent of the region of occipital cortex that when stimulated gives phosphenes fixed in the visual field. *J. Physiol.*, 225:57–58, 1972.

[45] G. S. Brindley. *Sensory effects of electrical stimulation of the visual and paravisual cortex in man*, volume vnI/3B of *Handbook of Sensory Physiology*. Springer-Verlag, New York, 1973.

[46] G. S. Brindley and P. E. K. Donaldson. Striate cortex stimulator. *US Patent #3,699,970*, Oct. 24, 1972.

[47] W. H. Dobelle and M. G. Mladejovsky. Phosphenes produced by electrical stimulation of human occipital cortex, and their application to the development of a prosthesis for the blind. *J. Physiol.*, 243(2):553–576, Dec. 1974.

[48] E. M. Schmidt, M. J. Bak, F. T. Hambrecht, C. V. Kufta, D. K. O'Rourke, and P. Vallabhanath. Feasibility of a visual prosthesis for the blind based on intracorticai microstimulation of the visual cortex. *Brain*, 119:507–522, 1996. DOI: 10.1093/brain/119.2.507

[49] W. H. Dobelle. Artificial vision for the blind by connecting a television camera to the visual cortex. *ASAIO J.*, 46(1):3–9, Jan. - Feb. 2000. DOI: 10.1097/00002480-200001000-00002

[50] R. A. Normann, E. M. Maynard, P. J. Rousche, and D. J. Warren. A neural interface for a cortical vision prosthesis. *Vision Res.*, 39(15):2577–2587, Jul. 1999. DOI: 10.1016/S0042-6989(99)00040-1

[51] E. J. Tehovnik and W. M. Slocum. Phosphene induction by microstimulation of macaque v1. *Brain Res. Rev.*, 53(2):337–343, Feb. 2007. DOI: 10.1016/j.brainresrev.2006.11.001

[52] J. J. Struijk, M. Thomsen, J. O. Larsen, and T. Sinkjaer. Cuff electrodes for long-term recording of natural sensory information. *IEEE Eng. Med. Biol. Mag.*, 18(3):91–98, May - Jun. 1999. DOI: 10.1109/51.765194

[53] G. G. Naples, J. T. Mortimer, A. Scheiner, and J. D. Sweeney. A spiral nerve cuff electrode for peripheral nerve stimulation. *IEEE Trans. Biomed. Eng.*, 35(11):905–916, Nov. 1988. DOI: 10.1109/10.8670

[54] M. Sahin, M. A. Haxhiu, D. M. Durand, and I. A. Dreshaj. Spiral nerve cuff electrode for recordings of respiratory output. *J. Appl. Physiol.*, 83(1):317–322, Jul. 1997.

[55] C. Veraart, C. Raftopoulos, J. T. Mortimer, J. Delbeke, D. Pins, G. Michaux, A. Vanlierde, S. Parrini, and M. C. Wanet-Defalque. Visual sensations produced by optic nerve stimulation using an implanted self-sizing spiral cuff electrode. *Brain Res.*, 813(1):181–186, Nov. 30, 1998. DOI: 10.1016/S0006-8993(98)00977-9

[56] J. Perez-Orive and D. M. Durand. Modeling study of peripheral nerve recording selectivity. *IEEE Trans. Rehabil. Eng.*, 8(3):320–329, Sept. 2000.

[57] D. J. Tyler and D. M. Durand. Functionally selective peripheral nerve stimulation with a flat interface nerve electrode. *IEEE Trans. Neural Syst. Rehabil. Eng.*, 10(4):294–303, Dec. 2002. DOI: 10.1109/TNSRE.2002.806840

[58] D. K. Leventhal and D. M. Durand. Subfascicle stimulation selectivity with the flat interface nerve electrode. *Ann. Biomed. Eng.*, 31(6):643–652, Jun. 2003. DOI: 10.1114/1.1569266

[59] A. F. Marks. Bullfrog nerve regeneration into porous implants. *Anatom. Rec.*, 162:226, 1969.

[60] R. Llinas, C. Nicholson, and K. Johnson. *Implantable monolithic wafer recording electrodes for neurophysiology*, Chapter 7, pp. 105–111. Brain Unit Activity During Behavior. Thomas, Springfield, IL, 1st ed., 1973.

[61] G. T. Kovacs, C. W. Storment, and J. M. Rosen. Regeneration microelectrode array for peripheral nerve recording and stimulation. *IEEE Trans. Biomed. Eng.*, 39(9):893–902, 1992. DOI: 10.1109/10.256422

[62] G. T. Kovacs, C. W. Storment, M. Halks-Miller, C. R. Belczynski, C. C. Della Santina, E. R. Lewis, and N. I. Maluf. Silicon-substrate microelectrode arrays for parallel recording of neural activity in peripheral and cranial nerves. *IEEE Trans. Biomed. Eng.*, 41(6):567–77, 1994. DOI: 10.1109/10.293244

[63] A. F. Mensinger, D. J. Anderson, J. Buchko, M. A. Johnson, D. C. Martin, P. A. Tresco, R. B. Silver, and S. M. Highstein. Chronic recording of regenerating viiith nerve axons with a sieve electrode. *J. Neurophys.*, 83(1):611–616, 2000.

[64] A. Ramachandran, M. Schuettler, N. Lago, T. Doerge, K. P. Koch, X. Navarro, K. P. Hoffmann, and T. Stieglitz. Design, in vitro and in vivo assessment of a multi-channel sieve electrode with integrated multiplexer. *J. Neural Eng.*, 3:114–124, 2006. DOI: 10.1088/1741-2560/3/2/005

[65] T. Kawada, C. Zheng, S. Tanabe, T. Uemura, K. Sunagawa, and M. Sugimachi. A sieve electrode as a potential autonomic neural interface for bionic medicine. *Proc. IEEE Eng. Med. Biol. Soc. 26th Annu. Int. Conf.*, 2:4318–4321, Sept. 2004.

[66] P. K. Campbell, K. E. Jones, and R. A. Normann. A 100 electrode intracortical array: structural variability. *Biomed. Sci. Instrum.*, 26:161–165, 1990.

[67] K. E. Jones, P. K. Campbell, and R. A. Normann. A glass/silicon composite intracortical electrode array. *Ann. Biomed. Eng.*, 20(4):423–437, 1992. DOI: 10.1007/BF02368134

[68] E. M. Maynard, C. T. Nordhausen, and R. A. Normann. The utah intracortical electrode array: a recording structure for potential brain-computer interfaces. *Electroencephalogr. Clin. Neurophysiol.*, 102(3):228–239, Mar. 1997.

[69] C. T. Nordhausen, E. M. Maynard, and R. A. Normann. Single unit recording capabilities of a 100 microelectrode array. *Brain Res.*, 726(1-2):129–140, Jul. 8, 1996. DOI: 10.1016/0006-8993(96)00321-6

[70] P. J. Rousche and R. A. Normann. Chronic intracortical microstimulation (icms) of cat sensory cortex using the utah intracortical electrode array. *IEEE Trans. Rehabil. Eng.*, 7(1):56–68, Mar. 1999. DOI: 10.1109/86.750552

[71] E. M. Maynard, N. G. Hatsopoulos, C. L. Ojakangas, B. D. Acuna, J. N. Sanes, R. A. Normann, and J. P. Donoghue. Neuronal interactions improve cortical population coding of movement direction. *J. Neurosci.*, 19(18):8083–8093, Sept. 15, 1999.

[72] J. M. R. Delgado. Permanent implantation of multilead electrodes in the brain. *Yale J. Biol. Med.*, 24:351–358, 1952.

[73] C. Palmer. A microwire technique for recording single neurons in unrestrained animals. *Brain Res. Bull.*, 3(3):285–289, May - Jun. 1978. DOI: 10.1016/0361-9230(78)90129-6

[74] G. W. Westby and H. Wang. A floating microwire technique for multichannel chronic neural recording and stimulation in the awake freely moving rat. *J. Neurosci. Meth.*, 76(2):123–133, Oct. 3, 1997. DOI: 10.1016/S0165-0270(97)00088-5

[75] J. G. Cham, E. A. Branchaud, Z. Nenadic, B. Greger, and J. W. Burdick. Semi-chronic motorized microdrive and control algorithm for autonomously isolating and maintaining optimal extracellular action potentials. *J. Neurophys.*, 93(1):570–579, Jan. 2005.

[76] C. Pang, Y. C. Tai, J. W. Burdick, and R. A. Andersen. Electrolysis-based diaphragm actuators. *Nanotechnology*, 17:S64–S68, 2006. DOI: 10.1088/0957-4484/17/4/010

[77] WIMS-ERC. Wims erc annual report 2007. Technical report, University of Michigan, Ann Arbor, 2007.

[78] K. D. Wise, A. M. Sodagar, Y Yao, M. N. Gulari, G. E. Perlin, and K. Najafi. Microelectrodes, microelectronics, and implantable neural microsystems. *Proc. IEEE, Special Issue on Implantable Biomimetic Microelectronics Systems*, Jul. 2008.

[79] G. E. Perlin and K. D. Wise. A compact for three-dimensional neural microelectrode arrays. *IEEE Int. Engr. In Med. Biol. Conf. Proc.*, Aug. 2008.

[80] J. P. Seymour and D. R. Kipke. Fabrication of polymer neural probes with sub-cellular features for reduced tissue encapsulation. *Conf. Proc. IEEE Eng. Med. Biol. Soc.*, 1:4606–4609, 2006. DOI: 10.1109/IEMBS.2006.260528

[81] Y. Kato, I. Saito, T. Hoshino, T. Suzuki, and K. Mabuchi. Preliminary study of multichannel flexible neural probes coated with hybrid biodegradable polymer. *Conf. Proc. IEEE Eng. Med. Biol. Soc.*, 1:660–663, 2006. DOI: 10.1109/IEMBS.2006.259978

[82] J. Wang, M. N. Gulari, and K. D. Wise. A parylene-silicon cochlear electrode array with integrated position sensors. *Conf. Proc. IEEE Eng. Med. Biol. Soc.*, 1:3170–3173, 2006.

[83] S. Kisban, S. Herwik, K. Seidl, B. Rubehn, A. Jezzini, M. A. Umiltà, L. Fogassi, T. Stieglitz, O. Paul, and P. Ruther. Microprobe array with low impedance electrodes and highly flexible polyimide cables for acute neural recording. *Conf. Proc. IEEE Eng. Med. Biol. Soc.*, 2007:175–178, 2007.

[84] E. Patrick, M. Ordonez, N. Alba, J. C. Sanchez, and T. Nishida. Design and fabrication of a flexible substrate microelectrode array for brain machine interfaces. *Conf. Proc. IEEE Eng. Med. Biol. Soc.*, 1:2966–2969, 2006. DOI: 10.1109/IEMBS.2006.260581

[85] P. T. Kissinger, J. B. Hart, and R. N. Adams. Voltammetry in brain tissue. *Brain Res.*, 55(1):209–213, 1973. DOI: 10.1016/0006-8993(73)90503-9

[86] A. J. Bard and L. R. Faulkner. *Electrochemical Methods: Fundamentals and Applications*. John Wiley & Sons Ltd., New York, 2nd ed., Dec. 18, 2000.

[87] P. Fromherz, H. Schaden, and T. Vetter. Guided outgrowth of leech neurons in culture. *Neurosci. Lett.*, 129:77–80, 1991. DOI: 10.1016/0304-3940(91)90724-8

[88] P. Fromherz and H. Schaden. Defined neuronal arborizations by guided outgrowth of leech neurons in culture. *Eur. J. Neurosc.*, 6:1500–1504, 1994. DOI: 10.1111/j.1460-9568.1994.tb01011.x

[89] A. A. Prinz and P. Fromherz. Electrical synapses by guided growth of cultured neurons from the snail lymnaea stagnalis. *Biol. Cybern.*, 82:L1–L5, 2000. DOI: 10.1007/PL00007969

[90] P. Fromherz and V. Gaede. Exclusive-OR function of single arborized neuron. *Biol. Cybern.*, 69:337–344, 1993. DOI: 10.1007/BF00203130

[91] P. Fromherz and A. Stett. Silicon-neuron junction: Capacitive stimulation of an individual neuron on a silicon chip. *Phys. Rev. Lett.*, 75(8):1670–1673, Aug. 21, 1995. DOI: 10.1103/PhysRevLett.75.1670

[92] A. Stett, B. Müller, and P. Fromherz. Two-way silicon-neuron interface by electrical induction. *Phys. Rev. E.*, 55(2):1779–1782, Feb. 1997. DOI: 10.1103/PhysRevE.55.1779

[93] G. Zeck and P. Fromherz. Noninvasive neuroelectronic interfacing with synaptically connected snail neurons immobilized on a semiconductor chip. *Proc. Natl. Acad. Sci. U S A*, 98(18):10457–10462, Aug. 28, 2001. DOI: 10.1073/pnas.181348698

[94] M. Voelker and P. Fromherz. Signal transmission from individual mammalian nerve cell to field-effect transistor. *Small*, 1(2):206–210, Feb. 2005. DOI: 10.1002/smll.200400077

[95] M. Hutzler, A. Lambacher, B. Eversmann, M. Jenkner, R. Thewes, and P. Fromherz. High-resolution multitransistor array recording of electrical field potentials in cultured brain slices. *J. Neurophysiol.*, 96:1638–1645, Sept. 2006. DOI: 10.1152/jn.00347.2006

[96] J. Pine. Recording action potentials from cultured neurons with extracellular microcircuit electrodes. *J. Neurosci. Meth.*, 2(1):19–31, Feb. 1980. DOI: 10.1016/0165-0270(80)90042-4

[97] W. G. Regehr, J. Pine, and D. B. Rutledge. A long-term in vitro silicon-based microelectrode-neuron connection. *IEEE Trans. Biomed. Eng.*, 35(12):1023–1032, Dec. 1988. DOI: 10.1109/10.8687

[98] Scanning Microscopy International. *Neural Transplant Staining with DiI and Vital Imaging by 2-Photon Laser-Scanning Microscopy*, P.O. Box 66507, AMF O'Hare (Chicago), IL 60666, Aug. 1995.

[99] J. Pine, M.P. Maher, S. Potter, Y.C. Tai, S. Tatic-Lucic, J. Wright, G. Buzsaki, and A. Bragin. A cultured neuron probe. *Proc. IEEE-EMBS 18th Ann. Mtg.*, 5(31):2133–2135, Oct. 31, Nov. 3, 1996.

[100] M. P. Maher, J. Pine, J. Wright, and Y. C. Tai. The neurochip: a new multielectrode device for stimulating and recording from cultured neurons. *J. Neurosci. Meth.*, 87:45–56, 1999. DOI: 10.1016/S0165-0270(98)00156-3

[101] A. Tooker, J. Erickson, G. Chow, Y. C. Tai, and J. Pine. Parylene neurocages for electrical stimulation on silicon and glass substrates. *28th Annu. Conf. IEEE Eng. Med. Biol. Soc.*, pp. 4322–4325, New York, 2006. IEEE-EMBS.

[102] J. R. Moffitt, Y. R. Chemla, S. B. Smith, and C. Bustamante. Recent advances in optical tweezers. *Annu. Rev. Biochem.*, Feb. 28, 2008. DOI: 10.1146/annurev.biochem.77.043007.090225

[103] H. Zhang and K. K. Lui. Optical tweezers for single cells. *J. R. Soc. Interface*, Apr. 1, 2008. DOI: 10.1098/rsif.2008.0052

[104] L. Mitchem and J. P. Reid. Optical manipulation and characterisation of aerosol particles using a single-beam gradient force optical trap. *Chem. Soc. Rev.*, 37(4):756–69, Apr. 2008. DOI: 10.1039/b609713h

[105] J. D. Wen, L. Lancaster, C. Hodges, A. C. Zeri, S. H. Yoshimura, H. F. Noller, C. Bustamante, and I. Tinoco. Following translation by single ribosomes one codon at a time. *Nature*, 452(7187):598–603, Apr. 3, 2008. DOI: 10.1038/nature06716

[106] M. L. Hayashi, S. Y. Choi, B. S. Rao, H. Y. Jung, H. K. Lee, D. Zhang, S. Chattarji, A. Kirkwood, and S. Tonegawa. Altered cortical synaptic morphology and impaired memory consolidation in forebrain- specific dominant-negative PAK transgenic mice. *Neuron*, 42:773–787, Jun. 2004. DOI: 10.1016/j.neuron.2004.05.003

[107] G. E. Loeb and R. Davoodi. The functional reanimation of paralyzed limbs. *IEEE Eng. Med. Biol. Mag.*, 24(5):45–51, 2005. DOI: 10.1109/MEMB.2005.1511499

[108] E. Chaplin. Functional neuromuscular stimulation for mobility in people with spinal cord injuries. The parastep i system. *J. Spinal Cord. Med.*, 19(2):99–105, Apr. 1996.

[109] G. E. Loeb, F. J. Richmond, and L. L. Baker. The bion devices: injectable interfaces with peripheral nerves and muscles. *Neurosurg. Focus*, 20(5):1–9 (E2), May 15, 2006.

[110] E. Niedermeyer and F. Lopes da Silva. *Electroencephalography: Basic Principles, Clinical Applications, and Related Fields.* Lippincott Williams & Wilkins, New York, 1st ed., 2004.

[111] G. Schalk, D. J. McFarland, T. Hinterberger, N. Birbaumer, and J.R. Wolpaw. BCI2000: a general-purpose brain-computer interface (BCI) system. *IEEE Trans. Biomed. Eng.*, 51:1034–1043, Jun. 2004. DOI: 10.1109/TBME.2004.827072

[112] K. R. Müller, M. Tangermann, G. Dornhege, M. Krauledat, G. Curio, and B. Blankertz. Machine learning for real-time single-trial EEG-analysis: from brain-computer interfacing to mental state monitoring. *J. Neurosci. Meth.*, 167:82–90, Jan. 2008. DOI: 10.1016/j.jneumeth.2007.09.022

[113] B. Blankertz, G. Dornhege, M. Krauledat, K. R. Müller, and G. Curio. The non-invasive Berlin Brain-Computer Interface: fast acquisition of effective performance in untrained subjects. *Neuroimage*, 37:539–550, Aug. 2007. DOI: 10.1016/j.neuroimage.2007.01.051

[114] S. Waldert, H. Preissl, E. Demandt, C. Braun, N. Birbaumer, A. Aertsen, and C. Mehring. Hand movement direction decoded from MEG and EEG. *J. Neurosci.*, 28:1000–1008, Jan. 2008. DOI: 10.1523/JNEUROSCI.5171-07.2008

[115] J.d.e.l. R. Millán, F. Renkens, J. Mouriño, and W. Gerstner. Noninvasive brain-actuated control of a mobile robot by human EEG. *IEEE Trans. Biomed. Eng.*, 51:1026–1033, Jun. 2004.

[116] Y. Washizawa, Y. Yamashita, T. Tanaka, and A. Cichocki. Extraction of steady state visually evoked potential signal and estimation of distribution map from EEG data. *Conf. Proc. IEEE Eng. Med. Biol. Soc.*, 2007:5449–5452, 2007.

[117] S. M. Sukthankar and N. P. Reddy. Virtual reality of "squeezing" using hand EMG controlled finite element models. In *Proc. of the 15th Annual International Conference of the IEEE*, Engineering in Medicine and Biology Society, pp. 972–972, 1993.

[118] Y. Barniv. Using electromyography to predict head motion for virtual reality. Technical report, National Aeronautics and Space Administration, Ames Research Center, 2001.

[119] L. J. Trejo, K. R. Wheeler, and C. C. Jorgensen. Multimodal neuroelectric interface development. Technical report, National Aeronautics and Space Administration, Ames Research Center, 2001.

[120] A. M. Sodagar, K. D. Wise, and K. Najafi. General telemetry chip for power and bidirectional data telemetry in implantable microsystems. *WIMS World*, p. 4, Summer 2007.

[121] M. T. Pardue, M. J. Phillips, H. Yin, B. D. Sippy, S. Webb-Wood, A. Y. Chow, and S. L. Ball. Neuroprotective effect of subretinal implants in the RCS rat. *Inv. Opthalmo. Vis. Sci.*, 46(2):674–682, Feb. 2005. DOI: 10.1167/iovs.04-0515

[122] T. Morimoto, T. Fujikado, J-S Choi, H. Kanda, T. Miyoshi, Y. Fukuda, and Y. Tano. Transcorneal electrical stimulation promotes the survival of photoreceptors and preserves retinal function in royal college of surgeons rats. *Inv. Opthalmo. Vis. Sci.*, 48(10):4725–4732, Oct. 2007. DOI: 10.1167/iovs.06-1404

[123] A. Y. Chow. Artificial retina device. *US Patent #5,016,633*, Aug. 8, 1989.

[124] G. Peyman, A. Y. Chow, C. Liang, V. Y. Chow, J. I. Perlman, and N. S. Peachey. Subretinal semiconductor microphotodiode array. *Ophthalmic Surg. Lasers*, 29:234–241, Mar. 1998.

[125] M. T. Pardue, S. L. Ball, J. R. Hetling, V. Y. Chow, A. Y. Chow, and N. S. Peachey. Visual evoked potentials to infrared stimulation in normal cats and rats. *Doc. Ophthalmol.*, 103:155–162, Sept. 2001. DOI: 10.1023/A:1012202410144

[126] A. Y. Chow, M. T. Pardue, J. I. Perlman, S. L. Ball, V. Y. Chow, J. R. Hetling, G. A. Peyman, C. Liang, E. B. Stubbs, and N. S. Peachey. Subretinal implantation of semiconductor-based photodiodes: durability of novel implant designs. *J. Rehabil. Res. Dev.*, 39:313–321, 2002.

[127] A. Y. Chow, V. Y. Chow, K. H. Packo, J. S. Pollack, G. A. Peyman, and R. Schuchard. The artificial silicon retina microchip for the treatment of vision loss from retinitis pigmentosa. *Arch. Ophthalmol.*, 122(4):460–469, 2004. DOI: 10.1001/archopht.122.4.460

[128] A. Y. Chow and V. Y. Chow. Subretinal electrical stimulation of the rabbit retina. *Neurosci. Lett.*, 225:13–16, Mar. 1997. DOI: 10.1016/S0304-3940(97)00185-7

[129] T. Morimoto, T. Miyoshi, T. Fujikado, Y. Tano, and Y. Fukuda. Electrical stimulation enhances the survival of axotomized retinal ganglion cells in vivo. *Neuroreport*, 13(2):227–230, Feb. 2002. DOI: 10.1097/00001756-200202110-00011

[130] P. J. DeMarco Jr., G. L. Yarbrough, C. W. Yee, G. Y. McLean, B. T. Sagdullaev, S. L. Ball, and M. A. McCall. Stimulation via a subretinally placed prosthetic elicits central activity and induces a trophic effect on visual responses. *Inv. Opthalmo. Vis. Sci.*, 48(2):916–926, Feb. 2007. DOI: 10.1167/iovs.06-0811

[131] E. Zrenner. [Initial success in retinitis pigmentosa patients. Retinal chip brings hope to the blind (interview by Dr. med. Thomas Meissner)]. *MMW Fortschr Med.*, 149:16, Apr. 2007.

[132] M. S. Humayun. Intraocular retinal prosthesis. *Trans. Amer. Ophthalmol. Soc.*, 99:271–300, 2001.

[133] R. R. Lakhanpal, D. Yanai, J. D. Weiland, G. Y. Fujii, S. Caffey, R. J. Greenberg, E. de Juan, and M. S. Humayun. Advances in the development of visual prostheses. *Curr. Opin. Ophthalmol.*, 14:122–127, Jun. 2003. DOI: 10.1097/00055735-200306000-00002

[134] M. Javaheri, D. S. Hahn, R. R. Lakhanpal, J. D. Weiland, and M. S. Humayun. Retinal prostheses for the blind. *Ann. Acad. Med. Singap.*, 35:137–144, Mar. 2006.

[135] J. D. Weiland and M. S. Humayun. Intraocular retinal prosthesis. Big steps to sight restoration. *IEEE Eng. Med. Biol. Mag.*, 25:60–66, 2006. DOI: 10.1109/MEMB.2006.1705748

[136] Cochlea implants: National institute on deafness and other communication disorders (nidcd), 2008. http://www.nidcd.nih.gov/health/hearing/coch.asp.

[137] J. E. Colgate and M. A. Peshkin. Passive robotic constraint devices using non-holonomic transmission elements. *US Patent #5,923,139*, Jul. 13, 1999.

[138] J. E. Colgate and M. A. Peshkin. Cobots. *US Patent #5,952,796*, Sept. 14, 1999.

[139] J. Horgan. The work of Jose Delgado, a pioneering star. *Sci. Amer.*, pp. 65–78, Oct. 2005.

Printed in the United States
by Baker & Taylor Publisher Services